Standards for Success

GIS for Federal Progress and Accountability

Christopher Thomas

ESRI Press
REDLANDS, CALIFORNIA

ESRI Press, 380 New York Street, Redlands, California 92373-8100

 First edition 2006
10 09 08 07 06   1 2 3 4 5 6 7 8 9 10

Printed in the United States of America

Library of Congress Cataloging-in-Publication Data
Thomas, Christopher–
Standards for success : GIS for federal progress and accountability / Christopher Thomas.—1st ed.
p. cm.
ISBN-13: 978-1-58948-047-6 (pbk. or hardcover : alk. paper)
ISBN-10: 1-58948-047-3 (pbk. or hardcover : alk. paper)
1. Geographic information systems—United States. 2. Administrative agencies—Information technology—United States. 3. Administrative agencies—United States—Decision making. I. Title.
G70.215.U6T46 2006
352.3'80285—dc22   2006037997

ISBN-13: 978-1-58948-087-2
ISBN-10: 1-58948-087-2

Ask for ESRI Press titles at your local bookstore or order by calling 1-800-447-9778. You can also shop online at www.esri.com/esripress. Outside the United States, contact your local ESRI distributor.

ESRI Press titles are distributed to the trade by the following:

In North America, South America, Asia, and Australia:
Ingram Publisher Services
Toll-free telephone: (800) 648-3104
Toll-free fax: (800) 838-1149
E-mail: customerservice@ingrampublisherservices.com

In the United Kingdom, Europe, and the Middle East:
Transatlantic Publishers Group Ltd.
Telephone: 44 20 7373 2515
Fax: 44 20 7244 1018
E-mail: richard@tpgltd.co.uk

Map cover by U.S. Geological Survey
Cover and interior design by Jennifer Jennings

Contents

Foreword

In 2004, ESRI released the highly successful *Measuring Up* volume that featured case studies showing how businesses and local governments achieve a return on their investments while realizing cost savings, improved efficiency in business processes, and enhanced decision support. In keeping with the spirit of that book's message, *Standards for Success* focuses on agencies of the federal government and how their best practices in GIS have led to successes in developing sustainable business-driven solutions.

The federal government is complex, and the managers within its agencies are constantly adjusting to the vicissitudes of growth and change while complying with standards for accountability and charting milestones for progress and performance. The common theme in *Standards for Success* is the implementation and integration of a successful GIS component that has helped administrators meet these challenges. Additionally, many agencies have found that by sharing geospatial data and processes across government teams, they realize improvements in business performance and services offered to citizens. As data is shared, its value increases, as does the influence of data analysis.

GIS software provides a new medium for understanding, modeling, and communicating the physical and cultural knowledge of our world. The technology excels at breaking down this information into components and providing us with a systematic and integrative framework for analysis and visualization. This appeals to the cognitive and intuitive ways we comprehend things.

GIS technology is evolving rapidly, and the Web is becoming the new platform for GIS where information is distributed in a collaborative environment and with immediacy. The latest works from GIS professionals, such as those featured in this book, are examples of how these efforts are making a difference not only within federal organizations, but also around the world.

—Jack Dangermond
President, ESRI

1

Streamlining Business Processes

Federal agencies are under enormous pressure to do more with less. They are expected to find ways to perform more efficiently, address the increasing demand for government services, and continue operating in a climate of shrinking budgets and amplified scrutiny. In addition, accountability has become the watchword in the federal sector. Policy makers have to support their decisions with evidence that they considered all aspects of an issue and that the benefits will reach the as many people as possible. Technology is helping to ease these challenges. Geographic information systems (GIS) software offers a reliable and integrated approach to more efficient, effective, and responsive public services.

Avoiding costs, saving time, and generating revenue add up to an efficient organization, as do streamlined workflows and decision processes. The flexibility of GIS software allows agencies to design enterprise-wide information management systems that contain a spatial component and integrate databases and applications from disparate systems. Agencies use geospatial applications to improve and streamline decision support. Common tasks that usually require a series of complex and time-consuming steps can be consolidated into easy-to-use automated tools that free personnel from repetitive processes and allow them to concentrate on creative solutions.

One of the greatest challenges facing federal agencies is the growing demand for government services. The Internet, coupled with server-based GIS software such as ArcIMS, provides a robust platform for delivering maps and other geographic solutions to an eager population of connected citizens, and technologies such as ArcWeb Services provide a cost-effective alternative to purchasing and maintaining large datasets. Distributed GIS computing gives agencies the opportunity to offer comprehensive data management and analysis functionality within the organization, while at the same time consolidating IT operations and centralizing application management.

Web-based applications have emerged as the optimal solution for reducing paperwork, automating tasks, and providing access to government databases. In fact, agencies are finding that early investments in building databases can pay off in terms of more effective and accurate responses to logistical and strategic problems. In addition, the federal government uses GIS for data gathering, usually combining on-the-ground collection activities with remotely sensed imagery and information to provide decision makers and the public with timely reports and maps that help them prepare for and recover from natural disasters and economic changes.

Crisis GIS: Preparing for and responding to volcanic eruptions in the United States

David W. Ramsey, Joel E. Robinson, Steve P. Schilling, Janet R. Schaeffer, and Frank A. Trusdell, Volcano Hazards Team, U.S. Geological Survey, http://volcanoes.usgs.gov/

The Volcano Hazards Program (VHP) of the U.S. Geological Survey (USGS) is charged with monitoring the nation's high-risk volcanoes and issuing timely warnings of potential volcanic hazards to emergency management authorities and to the affected population. GIS technology is a key component of VHP monitoring and response capabilities. GIS is used to produce electronic and paper maps illustrating volcano hazards and potentially threatened areas; visualize and analyze changes in the surface and volume of volcanic features; aid in geologic mapping; and catalog the locations, types, and specifications of remote monitoring instruments.

VHP GIS specialists have developed a Crisis GIS plan to ensure rapid, reliable delivery of spatial information for data analysis and visualization for any required location during a volcanic crisis within the United States. This plan is being implemented in preparation for future events and in response to continuing eruptions at Mount St. Helens, Washington; Kilauea Volcano, Hawaii; Augustine Volcano, Alaska; and Anatahan Volcano, Commonwealth of the Northern Mariana Islands (a U.S. commonwealth).

Criteria for crisis management

A well-designed and well-constructed GIS of U.S. volcano data is invaluable during a crisis, when timely analysis and communication of volcanic hazards are most critical. An effective Crisis GIS needs to meet several criteria. First, it must have the capacity to store multiple large datasets, including base layer data, elevation data, geologic maps, hazard assessment maps, satellite and remotely sensed data, aerial photography, and monitoring instrument locations for volcanoes throughout the United States.

Second, the database must be able to handle large tables of attributes. Datasets of monitoring instruments that provide data on surface change and the underground movement of magma can be very large because each record may have many attributes,

Figure 1.1 Seismic-station locations for Mount St. Helens, Washington, and vicinity. One function of Crisis GIS is to catalog the locations, types, and specifications of remote monitoring instruments. Red lines are major publicly accessible roads. The Johnston Ridge Observatory (JRO) station is the visitor center closest to the volcano.

Courtesy of USGS.

such as location, instrument type, manufacturer, owner, power source, and telemetry specifications. Monitoring instruments include sensors such as seismic stations that record earthquakes, GPS receivers that record ground deformation, tiltmeters that record ground inflation or deflation, and remote cameras.

Third, adding new datasets to the system must be efficient and straightforward, as new data is constantly being generated during volcano monitoring. Fourth, datasets must be easily updatable, as volcanoes can change rapidly during eruption, necessitating changes to hazard maps and possibly to the locations of monitoring instruments that are moved, added, removed, or even destroyed. Fifth, the database needs to be easy to organize, manage, and maintain, with minimal system administration required.

The system also must be readily accessible by VHP GIS specialists stationed at the Alaska Volcano Observatory (Anchorage and Fairbanks), the Cascades Volcano Observatory (Vancouver, Washington), the Hawaiian Volcano Observatory (Hawaii National Park), the Long Valley Observatory (Menlo Park, California), the Yellowstone Volcano Observatory (Menlo Park, California), USGS headquarters (Reston, Virginia), and the Smithsonian Institution (Washington, D.C.). Finally, establishing such a GIS must be economically viable.

Figure 1.2 Lidar and MASTER images of the crater of Mount St. Helens, Washington. The shaded-relief image on the left is from a lidar DEM from November 2003, before the last eruption began; the shaded-relief image in the middle is from a lidar DEM; the image on the right is from MASTER data from October 14, 2004, showing a growing, hot lava dome. The new lava dome was approximately 400 meters wide at that time. DEMs and remotely sensed data are used to visualize and analyze changes in the surface and volume of volcanic features.

Courtesy of USGS.

Implementing a crisis GIS

To meet these challenges, the ArcSDE database engine and SQL Server relational database were selected to support Crisis GIS. ArcSDE and SQL Server can manage raster and vector datasets that are much larger in file size and spatial extent than can be stored with typical "file and folder" architecture. With tools such as image pyramids and data compression, large datasets are quickly displayed on screen. Adding and updating datasets and attribute tables are simple through the ArcCatalog and ArcMap software interfaces. System administration and database management of ArcSDE and SQL Server are not overly time-consuming, and accessing the data stored on an ArcSDE server in one location from ArcGIS software clients at different locations is also quite easy, depending on the speed and security protocols of the remote client's network. Servers with fast processors, large quantities of RAM, and terabytes of high-speed disk space to support ArcSDE and SQL Server are affordable.

VHP GIS specialists began construction of Crisis GIS by assembling large raster datasets, including mosaicked digital elevation models (DEMs) and shaded relief maps at both 10- and 30-meter resolutions for most of the western United States, as well as raster catalogs of USGS 1:24,000-scale topographic maps for several western states. Spatially extensive vector datasets have been added to the system, such as political boundaries, hydrography coverage, transportation networks, a 1:24,000-scale quadrangle index, some geologic mapping, monitoring instrument locations, and federal land management boundaries for the nation. As of this writing, efforts are focused on adding more volcano-specific data to the database, such as hazard maps, more geologic maps, volcanic vent locations, and recent, high-resolution DEMs of currently erupting Mount St. Helens. The ArcSDE infrastructure of Crisis GIS is being expanded to store and serve this data with the purchase of an additional server and a quadrupling of available disk space.

Handling data efficiently

Initial results of the Crisis GIS implementation have been quite favorable. The system houses most of the raw materials needed for map creation—not only for crisis response, but also for day-to-day geologic investigation and hazard analysis of volcanoes not yet erupting. This leads to more efficient map production with a single place to search for data rather than many systems that may not be easily accessible. As more and more data is added, the role of the database as central data storehouse will expand, changing from primarily a catalog of base layer data to a working server for more dynamic datasets.

Figure 1.3 Geologic map of Mount Shasta, California. Geologic mapping will be a major component of the Crisis GIS database, as it is vital to volcano hazards assessment. Pink and orange lobes north of the mountain's summit are young lava flows.

Courtesy of Robert L. Christiansen, USGS.

The growing database is organized so that datasets can be easily found, saving time. Although still evolving, some standard naming conventions have developed as data is added to the database, also making the data easier to find in a timely manner. By sharing data between VHP observatories and offices, Crisis GIS helps eliminate duplication of effort in acquiring and storing data, saving both time and money. The only notable problem with the system has been its slow response times between servers in California and clients in Alaska and Hawaii. This issue should be resolved as network connections are improved for speed. If sufficient improvements cannot be made to the network, another possibility would be placing ArcSDE servers in Alaska and Hawaii and mirroring their content with other VHP servers, thereby having the data available everywhere. Having mirrored servers could be prudent, as network connectivity could be adversely affected by an eruption.

As a response tool, Crisis GIS enhances VHP ability to visualize and analyze volcano hazards and to quickly communicate threat-related information to emergency managers, the media, and the public. The system provides a stable, central data storehouse for VHP GIS specialists at different locations to organize, catalog, and share spatial data, saving time and money.

Future plans for Crisis GIS call for volcano monitoring data from remote instruments to be linked in real time with the database. The integration of monitoring data with cartographic spatial data in GIS could greatly enhance the utility of GIS as a situational awareness and decision-making tool for volcanic unrest and eruption. The Crisis GIS database architecture may also be expanded to store international volcano hazard assessments and geologic maps in support of the Volcano Disaster Assistance Program.

By sharing data between VHP observatories and offices, Crisis GIS helps eliminate duplication of effort in acquiring and storing data, saving both time and money.

Prioritizing livestock waste management funding with Web-based GIS

Jeri Sue Ponder, Randy Wiggins, Cheryl Grapes, and Michele Basham May, USDA Natural Resources and Conservation Service, http://www.wy.nrcs.usda.gov/programs/eqip/eqip.html

Historically, agriculture has been an important component of Wyoming's economic identity, and it is an essential part of the state's culture and lifestyle. One of Wyoming's main agricultural commodities is livestock, and the U.S. Department of Agriculture (USDA) works closely with this industry to provide assistance in a variety of ways.

In 2004, the USDA Natural Resources Conservation Service (NRCS), National Cartography and Geospatial Center (NCGC), and the Wyoming NRCS began work on a pilot project using ESRI ArcIMS software that was designed to assist evaluators in field offices statewide with prioritizing rancher and farmer applications for financial and technical assistance.

The USDA Environmental Quality Incentives Program (EQIP) is a voluntary Farm Bill conservation program for farmers and ranchers that promotes agricultural production and environmental quality as compatible national goals. EQIP offers financial and technical assistance to eligible participants to install or implement structural and management practices on eligible agricultural land. Producers can apply for cost sharing for many types of conservation practices. Demand for EQIP contracts often exceeds funding availability, and applications are evaluated and selected for funding based on locally developed procedures to optimize environmental benefits. Complying with confined-animal feeding operations (AFO/CAFO) regulations is one of the top natural resource concerns facing Wyoming producers.

In prior years, evaluating and ranking an EQIP application for funding was done mostly on paper and could take between two and four hours. Not only was the process time-consuming, it could also be somewhat subjective, as it took place in twenty-nine field offices across the state. Because the ranking procedure included geospatial decision making, a customized ArcIMS tool was piloted to help prioritize EQIP funding applications for livestock waste management. The project was designed to speed up the EQIP application evaluation process, to automatically rank the applications, and to provide greater objectivity.

Automating a manual process

The old ranking process was done on paper. One of the most potentially inaccurate and time-consuming tasks associated with evaluating a request for EQIP cost-share funding for livestock waste management assistance was locating the AFO/CAFO on paper maps or in a GIS and determining its proximity to water features, whether the operation occurred within an impacted watershed, and whether the operation overlay an impaired aquifer.

Each field office required hard copies of the EQIP application ranking criteria worksheet, along with an aerial photo of each operation. The evaluator then used ESRI ArcView software to add the GIS layers for impaired water bodies and aquifer sensitivity. (Finding the correct GIS layer or hard copy sometimes was a lengthy procedure.) Multiple geospatial processing steps were required to calculate a score for each funding request. After determining the location and identifying the attribute for that area, the evaluator selected the ranking score for that criterion and wrote it on the worksheet.

Each criterion was scored and added manually for the total ranking score. This process also included calculating the cost per animal unit and using a graph to determine those points. Opportunity for error presented itself with each step. The entire process took approximately two hours per application. After the ranking worksheet was completed, it was photocopied and mailed to the state office to be compiled with all the applications across the state. The county name, application number, total ranking points, and estimated cost-share dollars were manually transferred from each application into a Microsoft Excel spreadsheet to determine priority ranking and cumulative cost share dollars requested. There was more opportunity for error with this procedure and even the potential to lose applications. This step generally required an entire day and involved a lot of paper handling. The NRCS state office then reviewed completed applications.

The customized ArcIMS site enables NRCS staff in offices around the state to quickly and easily locate and submit a point on a map representing an existing livestock waste facility, automate the derivation of necessary spatial information, collect user input about a client's requirements via a forms page, store the data in a centralized database, and score and iteratively rank all submitted funding applications against one another.

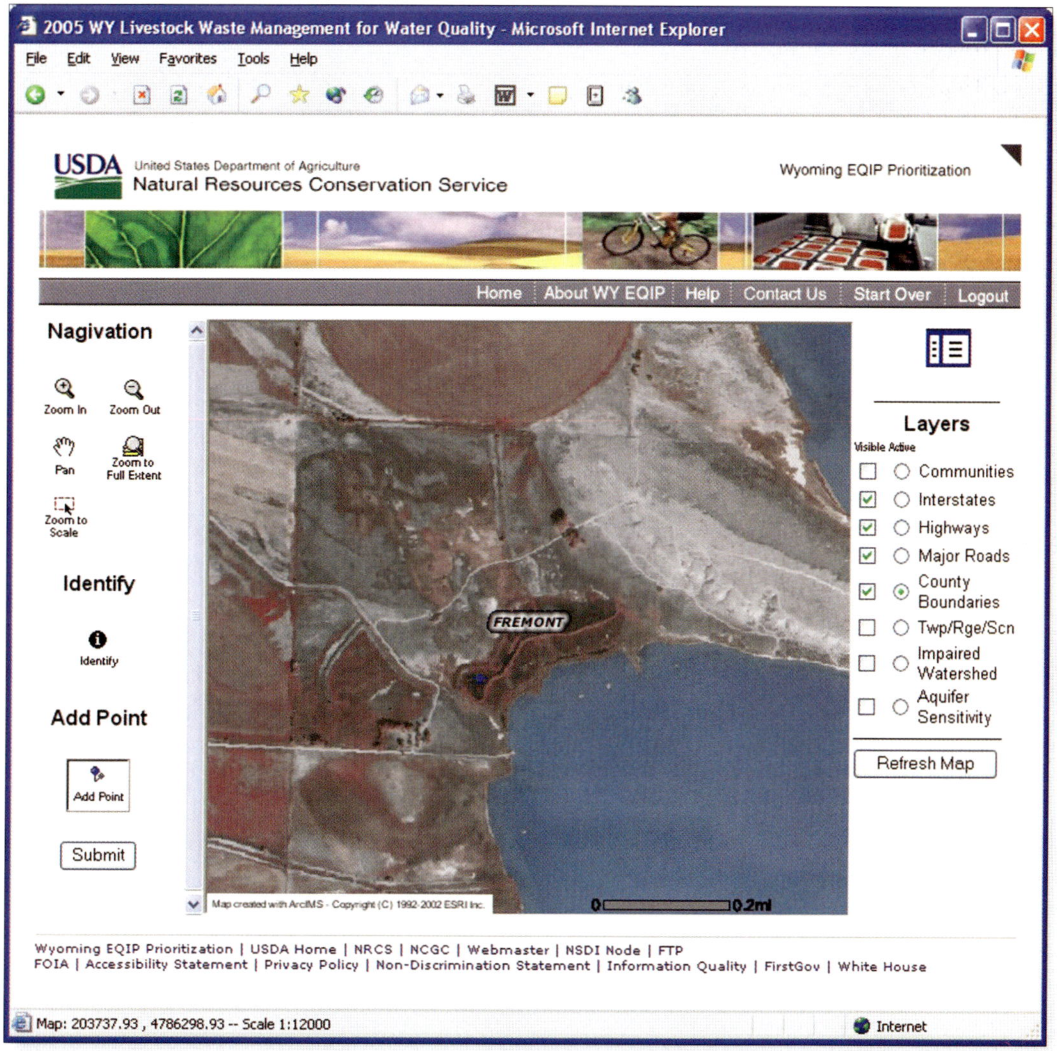

Figure 1.4 The Add Point tool is used to locate a new or existing AFO/CAFO on a Digital Orthophoto Quadrangle or Digital Raster Graphic at a predefined scale.

Courtesy of USDA-NRCS.

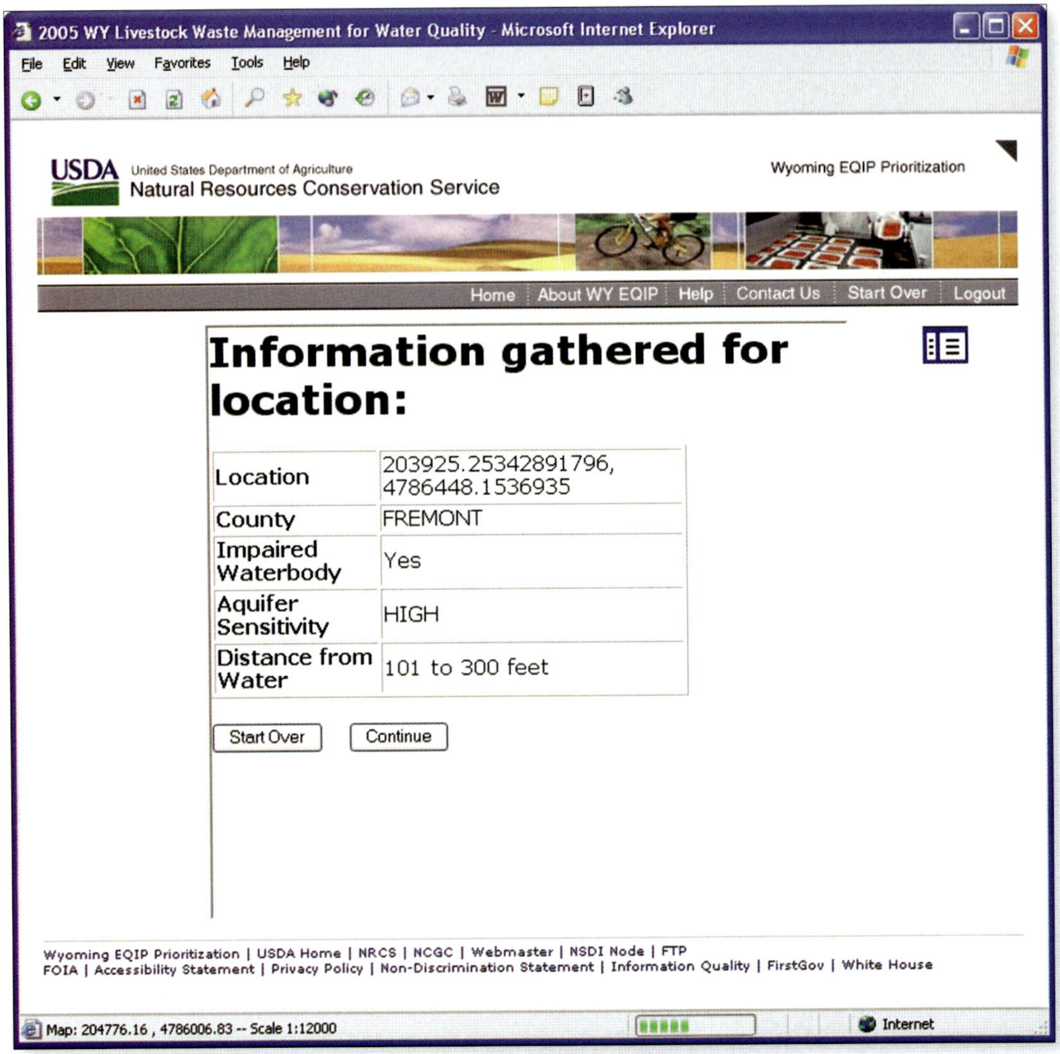

Figure 1.5 Overlays are used to determine whether an AFO/CAFO point falls within an impaired watershed or near a sensitive aquifer and how far it is from a water feature. The user has the option of relocating the point or submitting the point and continuing on with the data entry process.

Courtesy of USDA-NRCS.

Users of the ArcIMS site first log in from within the USDA network and then decide whether to add, view, or delete an EQIP application for livestock waste management funding. To add a new application, the user is forwarded to the ArcIMS site, where he can navigate and zoom to the location of the AFO/CAFO and then click on the map to place a point representing the operation via the Add Point tool. If the location is not correct, the user simply adds a point again. Only one facility is allowed per funding application. If the scale is 1:12,000 or better, the point is accepted, and a pushpin icon appears on the map. A message is displayed if the point is placed at a scale above 1:12,000, and the Zoom to Scale tool is provided to assist the user and ensure that all points are submitted at a consistent scale.

After the point is placed at the desired location and the Submit button is clicked, ArcIMS begins calculating spatial overlays in the background. When the overlays are complete, a table displays the point coordinates and the results of the overlay operations. The user is given the option of continuing or starting over. Next, a set of questions is presented via HTML forms.

When the data is submitted, a cost-effectiveness score is calculated, including total project cost, the cost-share amount, and the total points to be used to rank the application. The user is given the option of altering the questionnaire responses or submitting the application as is. User input is collected and stored along with the point coordinates and overlay results in an Oracle database table and used to calculate a score for the funding application. With proper permissions, a user may also view a list of existing applications or a table that ranks application scores and calculates cumulative cost-share funding requests for a county or state.

System configuration

The system uses ArcIMS 9 and runs on a Windows 2003 server with four processors and 2.0 GB of RAM. The Web server is Apache 2.0.48, and the servlet connector is Tomcat 4.129. Several Web programming languages were used, the combination of which provided a secure and robust application. The client is a modified HTML viewer with tools and forms written in JavaScript and HTML code. JavaServer Pages technology was used to wrap around the viewer for database connections, security, and session and variable management. Oracle 9.2 running under Solaris 9 software on a Sun 480 workstation serves as the RDBMS.

Data layers

The initial map view presents a painted relief image of the state of Wyoming along with county boundaries and major roads. As a user zooms in, additional vector layers are rendered, including the public land survey system, hydrography, and more detailed roads. Aquifer sensitivity data, classified from high to low, is also available for display. In addition, a watershed boundary layer showing impaired status is included.

ArcIMS performs three separate overlays after the point locating the livestock operation is submitted. The first overlay determines whether the point falls within an impaired watershed. The second overlay determines which of three proximity-to-water classes the point falls within. The last overlay uses the aquifer sensitivity layer to determine the aquifer sensitivity class in which the point falls. The resulting information is shown to the user on a new page, but the data cannot be changed. The user does have the option of returning to the map and changing the point location or of accepting the information and proceeding. The information acquired from the overlays is used in computing the final rank of the EQIP application.

Collecting EQIP applicant information

After a point locating the livestock operation is submitted, additional information about the operation must be collected. The priority of the application, the number of animals involved, and the total cost of the improvement are all needed to calculate the final ranking. The priority is determined by the current condition of the AFO/CAFO (high, medium, or low, based on environmental impact) and whether the operation will be removed if the proposed project is complete. The user must also indicate whether the applicant has completed a Self-Assessment Worksheet to determine his or her eligibility. Finally, the total cost of the project and the number of animals are needed to calculate cost-effectiveness. The answers to each of these questions are then shown to the user for review. The user can then submit his responses for calculating a score or return to the data input page to make changes.

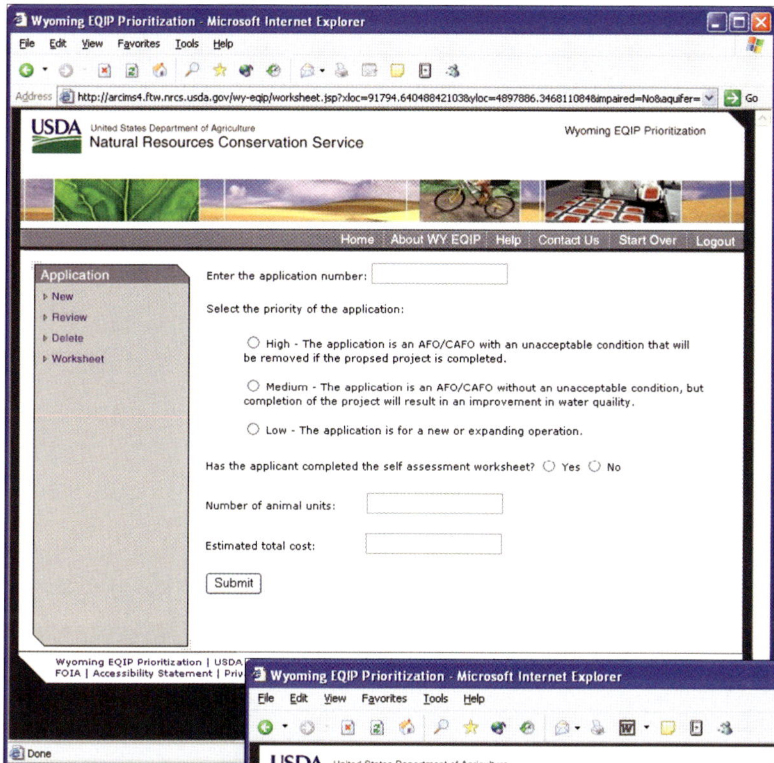

Figure 1.6
Information about the AFO/CAFO operation and the EQIP application for funding is collected via HTML forms and used to score and rank the application.

Courtesy of USDA-NRCS.

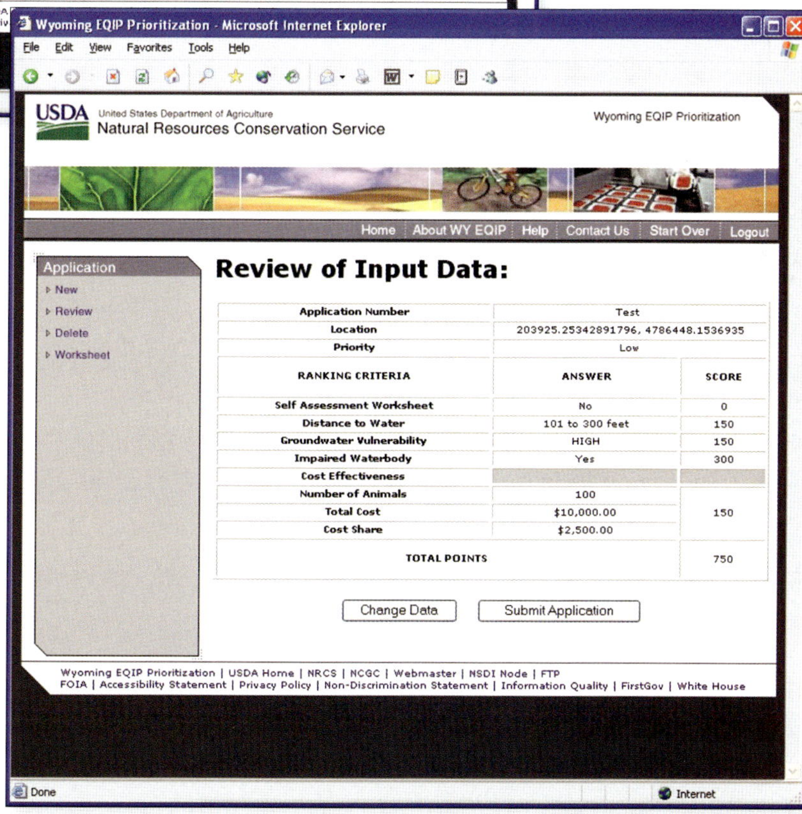

Figure 1.7
Responses to the EQIP questionnaire are displayed, and the user is given the opportunity to reenter the information or submit the application.

Courtesy of USDA-NRCS.

Calculating rank

When all the needed information is collected, the rank can be computed. This is done by assigning a predetermined number of points based on each response described above, in combination with the scores obtained from the geospatial overlays. These points are added to compute the final rank. Negative results are not allowed and are assigned a score of zero. A cumulative total for requested cost-share funds is also provided for the EQIP program administrator. After the rank is computed, the collected data results and rankings are stored in an Oracle database. This allows the state to review all submitted information and to create reports as desired. The NRCS state office uses a combination of rank and priority to determine which projects get funded for that year.

Security and administration

The application is housed inside the USDA firewall and is accessible only from within the USDA network. Users are assigned an ID and a password, and the USDA eAuthentication process handles user verification. Currently, two levels of user permissions are used: (1) field office users can add, delete, and review applications only for their county or service center area; (2) an administrator in the state office can add, delete, and review applications from across the state. A third level of user permissions for regional managers might be added in the future. Database and ArcIMS administration is handled by NCGC staff, while the state office is responsible for EQIP-related questions and general end-user support.

Several enhancements are planned. A tool to allow the state administrator to export the point coordinates, user responses, and all scores to an Excel spreadsheet or shapefile is under development. The application will also be modified to allow users with appropriate permissions to edit existing records without having to reenter the AFO/CAFO location and all other data. Where available, links to metadata will be added for each layer, and the legend and layer list will be integrated. The ability to zoom to a location based on township, section, and range values may also be implemented. Finally, brief Help documentation will be added.

Saving time and increasing productivity

Previously, ranking worksheets could take between two and four hours per application to complete. The new process takes just a few minutes with the ArcIMS tool. The field office staff need not open other references, find maps, or handle or make photocopies. In addition, all of the applications are already compiled, prioritized, and ranked for the state office, along with the cumulative dollars requested. The user can see how the completed application ranks against other applications submitted in their county or service center area. Selection of applications by the state office can be done almost immediately based on the aggregated information. This is a major time saver for state and field office personnel.

The advantages of the new system include a large savings of time and the increased accuracy and consistency of the ranking process. In the future, the tool could be made publicly available on the Web so that cost-share applicants could complete the ranking themselves from their home computer or from a kiosk at their local office. ArcIMS is a highly customizable and flexible program that can integrate geospatial questions with questions for the applicant to answer. It took approximately two months, plus the time to coordinate between NCGC and Wyoming NRCS personnel, to develop the system. Because the development process and programming code are applicable and transferable to other Web-enabled GIS prioritization tools, they can help reduce the time that would be required to develop similar tools.

Previously, ranking worksheets could take between two and four hours per application to complete. The new process takes just a few minutes with the ArcIMS tool.

Online Maps Add to Best Practices

Internet mapping technologies add geographic data and analysis to a multitude of applications that serve organizations and communities with a variety of information ranging from environmental data to demographic reports. Increasingly, government agencies ranging from the National Oceanic and Atmospheric Administration to the U.S. Environmental Protection Agency to the U.S. Geological Survey are making spatial data available online via Internet mapping technology. Distributing geographic data and applications on the Internet provides opportunities to eliminate paperwork, make government information accessible to more people, and apply data from within an organization while integrating information from other agencies.

Three Web sites featured in this book have crossover benefits which concurrently add to their success in managing resources, effectively communicating, streamlining business processes, and making informed decisions. The NCES Web site (http://nces.ed.gov/surveys/sdds) quickly gives school districts access to vital information. The EQIP application helps the U.S. Department of Agriculture streamline a process and helps staff in field offices make decisions based on accurate information. Land-use managers now have a single point of access to current soil survey data at the USDA Web Soil Survey (http://soils.usda.gov/survey).

Dealing with the mundane in innovative ways

Sean Sinclair, Robert Houston, and James A. Robertson, U.S. Air Force

Trash is a fact of life that many take for granted. Generally, trash is placed in a container that is carefully positioned and then transported to a destination or sent to a processing facility. Some communities allow (or require) trash to be segregated from the recyclables. The Air Force is one such community. In performing its mission to protect the United States from enemies, the Air Force strives to comply with environmental regulations. This is no small feat; however, dedicated environmental managers and GIS professionals can provide simple and effective technological solutions.

The Environmental Flight unit at Dyess Air Force Base (near Abilene, Texas) is responsible for environmental regulation compliance, pollution prevention, natural and cultural resource preservation, and installation restoration. For many years, the unit has been a leader in waste and recycling issues for the U.S. Air Force and Air Combat Command. Managing trash/recycling containers had been a problematic issue from time to time at Dyess. Trash/recycling container surveys were required annually to record

Figure 1.8 Collecting trash/recycling container data.

Courtesy of USAF.

locations and problem areas. Surveys were completed manually using vehicles, personnel, and hand-written notes. Many issues and discrepancies arose from using this method, as exact locations were difficult to record and data from multiple collection efforts required consolidation. During the two-week inventory, containers were discovered to have been relocated or removed, making the data collection process cumbersome and resulting in inaccuracies. The need to reduce labor costs, save money, and minimize human error and miscommunication led to a search for a more efficient method of data collection.

Developing a GIS-based tracking system

Environmental Flight turned to the Dyess GIS office (GeoBase) for help in tracking and managing the trash/recycling container inventory. A team of professionals gathered to assess the issues and devise solutions. The following requirements were identified:

- Quick and easy data collection
- Customizable data collection attributes
- Attribute choices (ease of drop-down lists and the option to type in attributes not listed)
- Addition and subtraction of data and attributes
- Trend and problem area analysis
- Direct compatibility (import/export) with GIS
- Tracking of physical conditions
- Easy relocation of features with GPS
- Easy integration of new data into an existing dataset
- Portability and lightweight
- Short learning curve for basic use

The GeoBase office had Trimble GeoXT GPS units with ESRI ArcPad software installed and available to use for the survey. The GeoXT/ArcPad system for rapid data collection proved to be easy to learn, requiring minimal training.

To simplify and streamline the field data collection process, the GeoBase team set out to design a custom form in ArcPad Application Builder. A list of attributes to collect for each container, and a list of choices for each attribute, needed to be developed. Pull-down

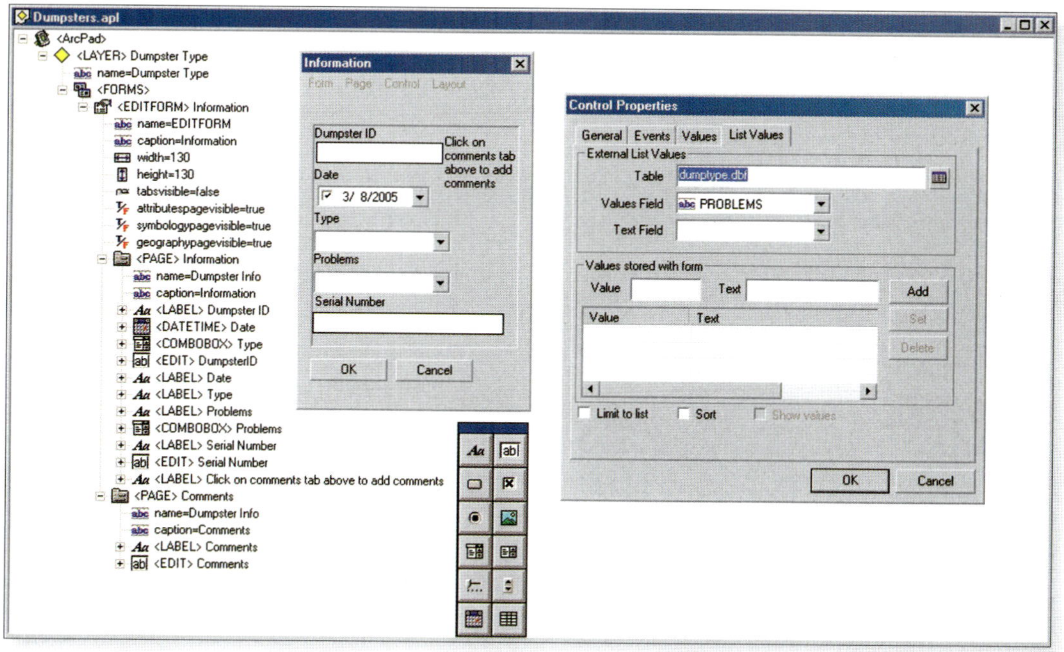

Figure 1.9 Linking the entry form to the attribute database with the ESRI Application Builder.

Data courtesy of USAF; ArcPad screen shot from ESRI.

menus, which save data collectors valuable time, were populated with the choices for each attribute. For unique situations, the option to manually enter a value for an attribute was added.

Application Builder incorporates convenient features such as multiple attributes and pull-down menus that reside in one table. Each part of the custom-developed form can be linked to the appropriate column in the table, and the choices for the attributes populate directly from the table. As the table is updated, the pull-down menu is also updated. Using one table for each form allows simple table development and eliminates cross-populating from multiple tables. For ease of use, the form is kept as brief as possible. Application Builder allows multiple page forms, but keeping the page count to a minimum enables users to quickly populate the required data.

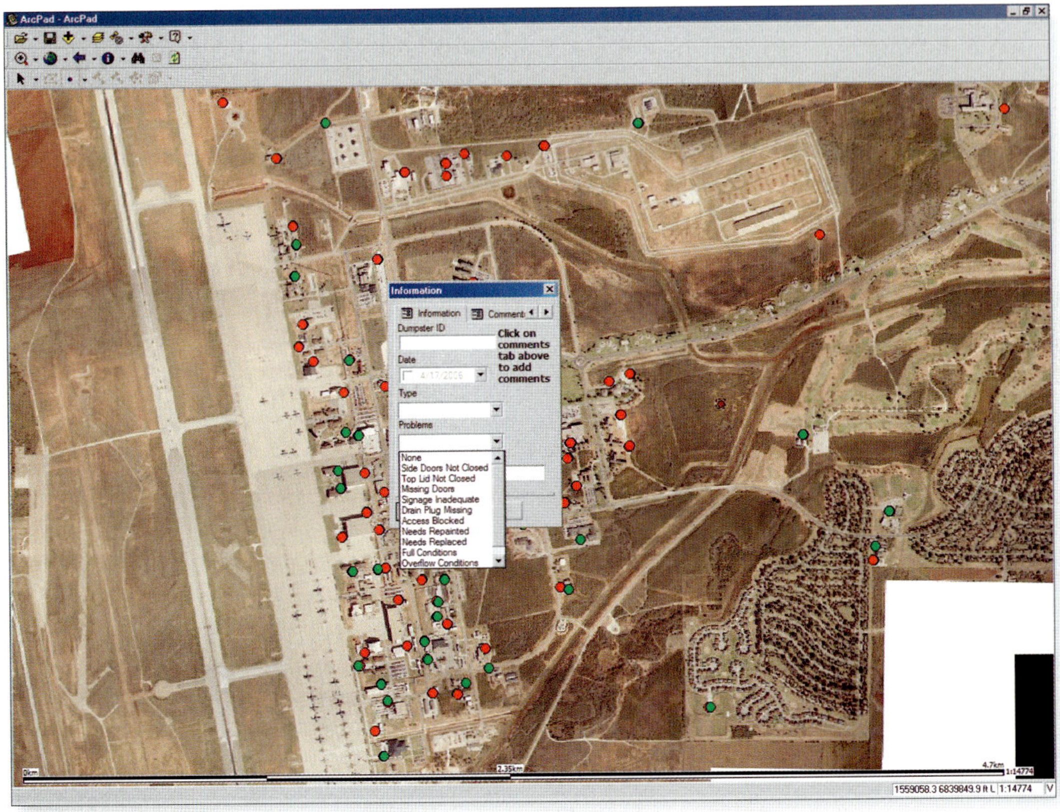

Figure 1.10 Entry form with the Problems drop-down list open.

Data courtesy of USAF; ArcPad screen shot from ESRI.

The new method for trash/recycling container surveying and tracking has resulted in a 90 percent savings in labor costs over previous methods.

The first page of the form contains all the vital information required. The first attribute, entered manually, is a container ID, which corresponds to the building number and ends in an "R" for recycling or a "T" for trash. The date and time a container is surveyed are recorded automatically from the Windows CE Pocket PC setting on the GeoXT device. The next attribute is container type—trash or recycle—which can be checked against the container ID and type collected in the field. Because only two choices are available for container type, a pull-down menu is used, eliminating the need to type in the words "Trash" or "Recycle." Other attributes are damage, deterioration, and serial number. A blank second page can be used for additional field notes. A shapefile is then created from all the entered attributes.

New tracking system in action

The trash/recycling container survey was completed in two days by one person. The collected data was returned to the office, and the shapefiles were downloaded. The data was then plotted on a map to determine contract compliance and areas for improvement.

The benefits of the new data collection method are many. The base has an accurate count of containers with their exact locations. The location information allowed Environmental Flight to identify issues such as poor container positioning or lack of containers in an area. Problems with specific containers were recorded, giving Environmental Flight the ability to ensure that repairs were made and damaged containers were replaced. Areas plagued by customer misuse were also recorded. Recyclable items found in trash containers, overflowing containers, and other misuses were highlighted during education and outreach opportunities. The data collection time savings, coupled with the easily updated container inventory and maintenance records, has made this project a winner for Dyess.

The new method for trash/recycling container surveying and tracking has resulted in a 90 percent savings in labor costs over previous methods. A more efficient allocation of resources and improved customer service were also realized. During a recent environmental audit, Dyess was rewarded for its efforts. The inspection team bestowed their highest honor for the new tracking system and recommended that this practice be implemented at other Air Force facilities.

Integrating technology with business processes for more effective management of naval facilities

Avinash Srivastava and Brad Wellington, EDAW, Inc.

Progress is often measured by lessons learned and experience gained as each milestone of a project is reached. For federal agencies, measuring progress highlights the importance of paying careful attention to existing business processes and workflows, integrating them into a new system design, and recognizing the benefits of an extended product lifecycle.

The U.S. Naval Surface Warfare Center (NSWC) Dahlgren Division in Virginia and the Patuxent River Naval Air Station (NAS) in Maryland needed an efficient, easy-to-operate accessible system for managing facilities and assets, with a centralized location for all datasets. The system would interface with the Naval Facility Assets Data Base, which would provide accurate and timely information for the annual General Services Administration real property inventory report. The two naval bases commissioned the design firm EDAW to create a facilities data viewer and asset management system.

Learning from existing systems

The naval base in Dahlgren has an approximate land area of 4,318 acres, with 736 facilities covering nearly 3.2 million square feet of floor space. The major tenants of the base include the NSWC Dahlgren Division, the AEGIS Training and Readiness Center, Joint Warfare Analysis Center, Naval Networks Space Operations Command, and the Surface Center for Combat Systems.

The Patuxent River NAS has an approximate base area of 6,420 acres, and its 787 facilities cover nearly 7.9 million square feet of floor area. The base's major tenants include the Naval Air Warfare Center-Aircraft Division, Naval Air Systems Command headquarters, and the U.S. Naval Test Pilot School.

At both sites, ArcView 3.x GIS-based customized facilities management (FM) solutions were in place—CADModule software for NAS at Patuxent River and Facilities Data Viewer (FDViewer) for NSWC at Dahlgren. Despite significant changes in GIS and CAD technology that have occurred since CADModule and FDViewer were implemented at the two sites in 2001, these systems are still in use. They were carefully integrated with business

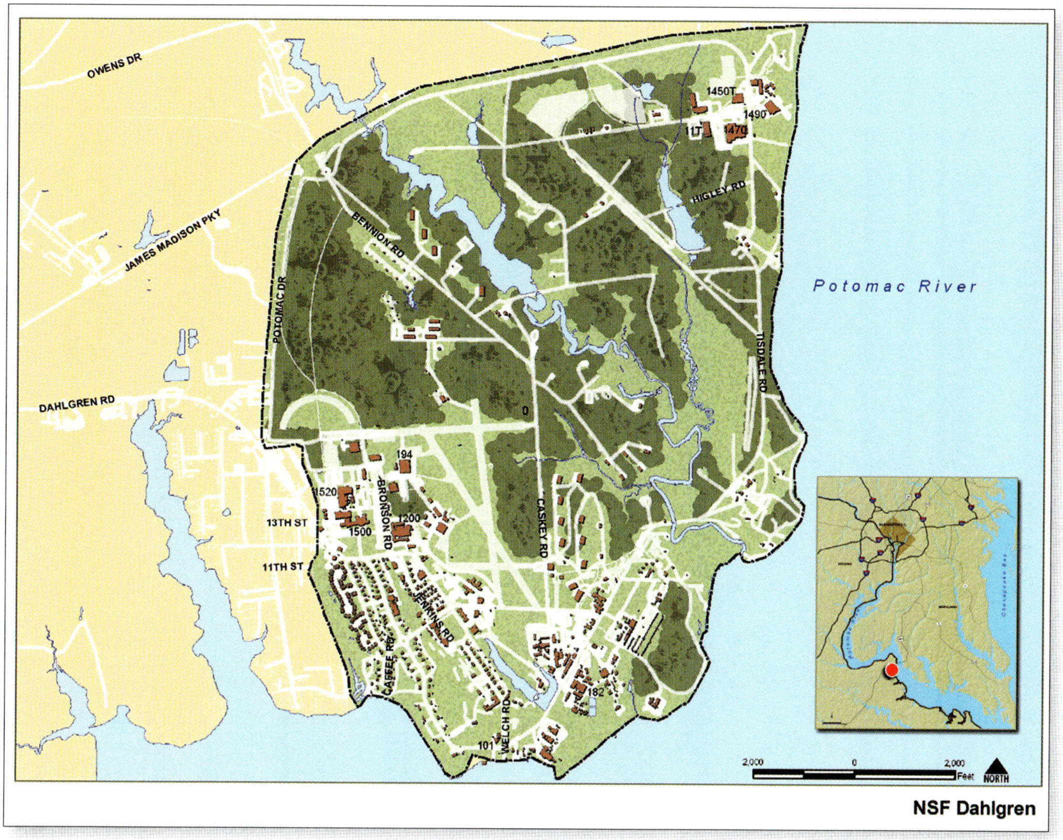

Figure 1.11 The GIS-based facility information management system at the U.S. Naval Surface Warfare Center (NSWC) Dahlgren Division in Virginia enables planners to create and maintain maps, floor plans, and databases related to facilities management.

Courtesy of U.S. Navy.

Integrating the new FIMS with existing databases and workflows was crucial for maximizing the return on investment and minimizing training and maintenance requirements.

processes and the needs of the users. Migration to the next generation system is underway, but the core functionality of the existing systems remains valid. A number of lessons can be learned from the success of these systems.

A successful facility information management system (FIMS) requires careful planning and a holistic understanding of the organization's mission, goals, structure, and key business processes. The development of the CADModule and FDViewer systems involved extensive user interviews identifying inputs, controls, outputs, mechanisms, needs, processes, and issues.

Key objectives in FIMS development

A diverse user group. The Public Works Department (PWD) at Dahlgren is the main user group dealing with facilities management, but users range from CAD/GIS technicians to planners and policy makers. The needs of the users range from emergency work orders at the room level, to space management and custodial assets at the facility level, to land planning issues at the base campus level. Approximately 20 percent of the end users are responsible for creating and maintaining maps, floor plans, and databases related to FM, while 80 percent require access to information for their decision making.

In the development of CADModule and FDViewer, great emphasis was placed on end user input. Joint application development sessions helped to define the basic graphic user interface structure and incorporate Navy-specific language. Working with user representatives helped minimize the learning curve and instill in the user group a sense of ownership of the application.

A number of key functions were defined, specifically tailored to the needs identified in the interview process and joint application development sessions.

- Seamless integration of base-level and floor plan information
- Simplified graphic user interface with multiple views
- Interactive information tools along with custom queries
- Integration of personnel, work management, and room information
- Basic metrics and reporting functions
- Interactive labeling and automated map layouts

Integration with existing systems. Integrating the new FIMS with existing databases and workflows was crucial for maximizing the return on investment and minimizing training and maintenance requirements. The Navy's Shore Facilities Planning System—outlining key business processes, data models, and information flows—provided important guidance.

In the case of FDViewer development, the challenge was to integrate databases and workflows from several departments. PWD Dahlgren, the main user group, had a number of information technology systems centered on the Public Works Management Automation System dealing with work order and emergency services information. In addition, the document management system FileNet was in place, and the Tri-Services standards for CAD and GIS data had been fully adopted as part of the basewide implementation of Citrix Winframe. AutoCAD and ArcView GIS were being used to create and update floor plans and basemaps. An extensive SQL server database with facilities, personnel, and organization-related information linked to the Naval Facilities Assets Data Base (now known as Internet Naval Facilities Assets Data Store) was also in use. A data migration plan was developed relatively easily for FDViewer.

Broad facility management application. Both Dahlgren and Patuxent River Naval facilities required their FIMS to incorporate an integrated lifecycle facilities management capability: integration of land and facilities data and seamless access to a variety of basemaps, building floor plans, and detailed room level information. The new system had to provide the following applications:

- Space management (current and proposed)
- Asset management and custodial services
- Work order and emergency service management
- Area development and master planning
- Data maintenance and verification

Because this range of applications was broader than the intended use of commercial, off-the-shelf, computer-aided facilities management systems, a custom FIMS was needed.

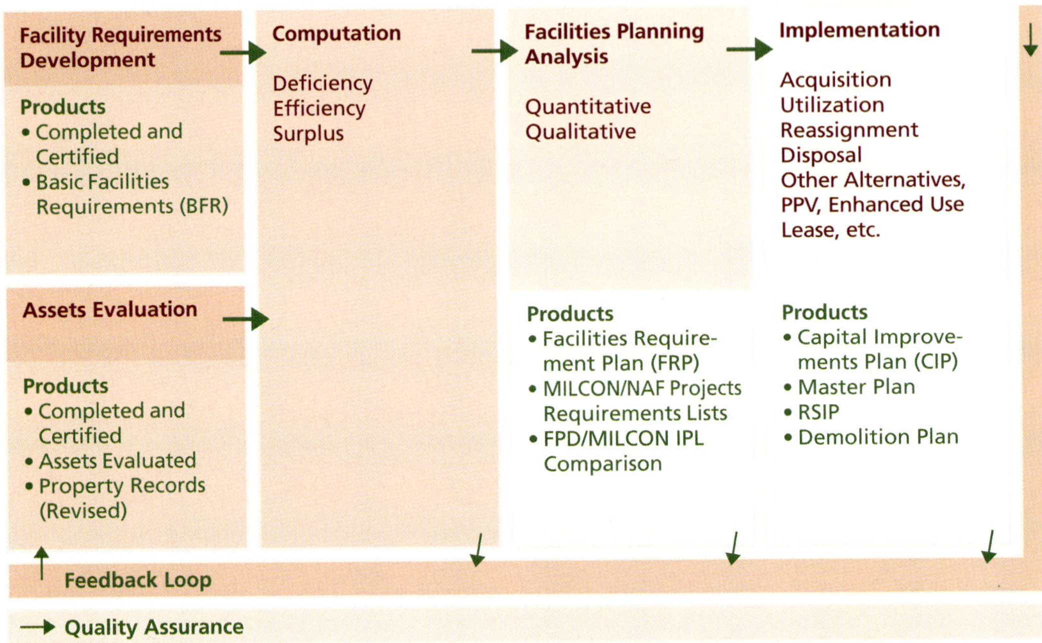

Figure 1.12 The Navy's Shore Facilities Planning System follows key business processes and information flows. Facilities, personnel, and organization-related information are linked to the Internet Naval Facilities Assets Data Store.

The right tool for the job

A GIS-based platform was chosen for the new FIMS rather than the traditional CAD platform. The display, analysis, and query capabilities of GIS technology integrate well with FM requirements, especially when CAD data is made easily accessible in a GIS environment. Both NSWC Dahlgren and NAS Patuxent River have comprehensive datasets in ArcInfo compatible formats, although GIS was traditionally used by a few specially trained GIS technicians.

The need to integrate land and facility information in a seamless application led to the selection of ArcView GIS software as the platform for CADModule and FDViewer. The main advantages of this approach were the ability to view, query, and edit both GIS and CAD data; a user-friendly interface easily customized with Avenue and Visual Basic programming languages; and the ability to link external databases (SQL server connections).

The two bases have approached new technological developments as an evolution rather than a revolution. The next generation of FDViewer will be Internet-based, allowing smart objects such as equipment, biomonitors, and sensors to transmit information directly to the FM databases. With GIS technology as its core, FDViewer has a solid foundation for expanding its functionality.

PWD Dahlgren has always been a leader in implementing information technology systems and has developed a new Web-based system using ArcIMS along with the Microsoft .NET Framework software. This prototype system incorporates the original FDViewer functionality and extends FM capabilities across all the installations in the Naval District's Washington Region. Similar functionality is also incorporated in the Navy's enterprise-wide RSIMS3 map viewer product.

The collection, review, analysis, and reporting of performance information is vital to FM managers and other decision makers. The new systems need to do more than just store facility information and provide a means to view it. They supply essential business metrics that relate to strategic organizational priorities. The next generation of FDViewer will need to provide benchmark comparisons for metrics such as space utilization per person or department; occupancy rates; cost per gross area; cost per building versus replacement costs; energy use per building, space type, or department; maintenance backlogs per building; and service call turn-around times.

FDViewer and CADModule successes were largely due to the fact that the EDAW development team's approach analyzed the business processes involved before implementing new technology. This resulted in increased duration of the application lifecycle. Lessons learned included:

- Consider business processes rather than a technology solution that forces a change in established practices.
- Use a risk-based approach to define requirements; don't aim for the ultimate do-all solution.
- Focus only on those tools and metrics that you really need.
- Take small bites and build from small successes.
- Don't forget the human factor. Established workflows and culture are important to success.
- Aim for evolution rather than revolution in lifecycle planning.

2

Fostering Collaboration and Communication

Federal agencies reap significant benefits in their business operations when they work within a framework of collaboration and communication on geospatial issues. The language of geography easily illustrates common issues and concerns and allows participants to see and understand each other's messages. That's why GIS software works as a tool for fostering teamwork and helping organizations build necessary, lasting relationships.

When federal partners work toward establishing a collaborative model for geospatial activities and investments, they ensure a sustainable and coordinated approach to producing, maintaining, and using geospatial data. When everyone is working from the same page, things run more smoothly. Collaborating agencies share information and data. They reorganize and standardize their workflows and implement common operating procedures so that they are interchangeable. They make their analyses available to each other and try to find universal solutions to mutual problems. In times of crisis, collaborating agencies (federal, state, and local) send their GIS personnel to incident sites so they can work together, blending technical skills and experience to help emergency managers make better decisions.

GIS is an effective tool for public participation, whereby governments can engage citizens in the public policy process. When issues are studied within a geographic context, relationships are more easily understood, and solutions are more discernible. Interaction between governments and their citizens helps define problems and clarify potential impacts of proposed projects. Network and Internet access to spatial databases hosted by government agencies helps to establish a rapport with the constituency.

Partnerships between federal agencies and private industries based on sharing geographic information are common. Private companies are contributing to government efforts to integrate legacy systems and to consolidate information technology architectures. These alliances enable federal agencies to be more innovative and react quickly to demands for government services while becoming leaner and more efficient.

Finally, GIS databases and applications serve as foundational tools for successful project development and deployment, which improve intra-agency functions by breaking down or transcending traditional informational "stovepipes" within departments. Information exchange and cooperative arrangements between agency departments are more readily achieved when GIS is integrated with standard operating procedures and decision making.

Web Soil Survey: A new way to access soil maps

Jim R. Fortner, National Soil Survey Center, http://websoilsurvey.nrcs.usda.gov/app/

The National Cooperative Soil Survey, a collaboration of federal, state, and local partners, recently made its soil survey data available by launching a GIS-enabled Web site known as Web Soil Survey (WSS).

Authorized by Congress in 1896 and placed under the auspices of the U.S. Department of Agriculture (USDA), the first soil survey field operations began in 1899. Today, the USDA Natural Resources Conservation Service (NRCS) oversees the National Cooperative Soil Survey (NCSS) program and manages the soil survey database for all NCSS partners. The soil survey is an inventory of the nation's soil resources designed to help land managers determine best uses and management practices for soils on their land parcels.

The survey uses a standard data model for collecting and distributing data, as well as standardized methods and procedures. Soil survey databases identify more than 300 soil properties, and by using digital imagery for overlaying data on data, they offer a tool for in-depth soil suitability analysis.

GIS helps soil survey data seekers to identify the best way to protect soil and water quality, as well as how land parcels are suited for specified land uses. GIS has been an important addition to the survey, adding capabilities for data access, analysis, spatial display, and now a high volume of distribution.

As of 2006, approximately 97 percent of privately owned U.S. land and 92 percent of all U.S. lands have been surveyed. Digitizing of soil survey maps began in the mid-1980s, with a concentrated effort beginning in 1997. As of March 2006, soil survey maps for approximately 2,500 soil survey areas (usually counties) have been digitized. The goal is to complete the digitizing of the approximately 500 remaining mapped survey areas by the end of 2007.

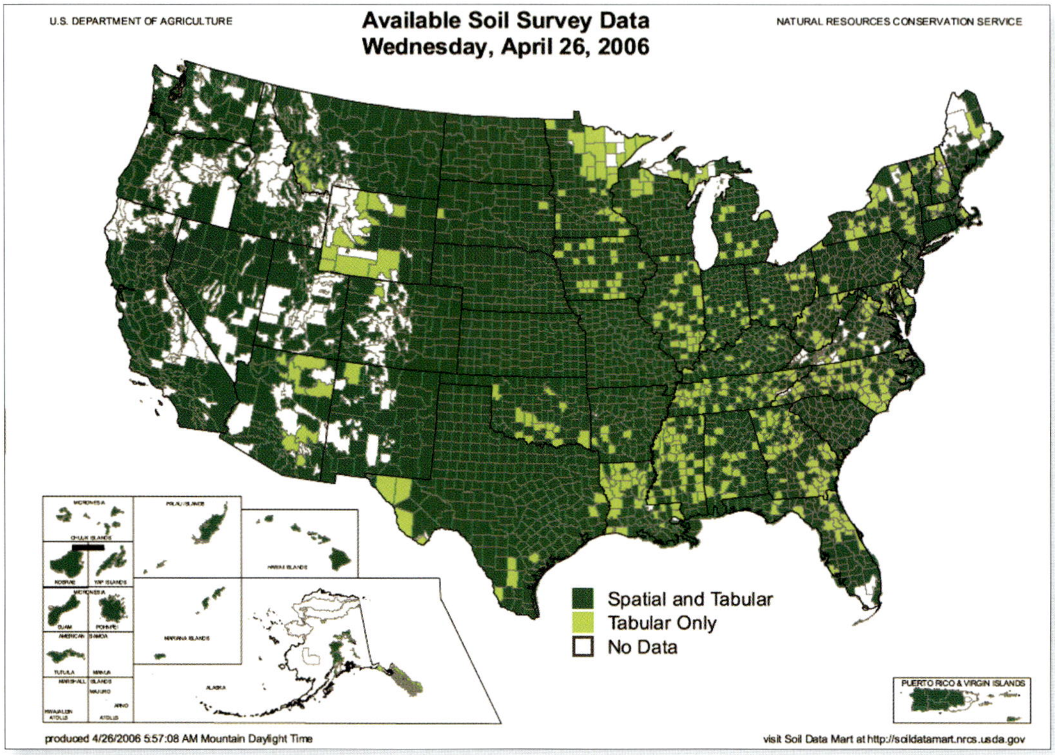

Figure 2.1 Map showing soil data availability. Areas shown in darker green have soil maps available for display in Web Soil Survey. Lighter green areas are mapped, but not digitized. This map is updated nightly as new data becomes available. The map can be accessed at http://websoilsurvey.nrcs.usda.gov/DataAvailability/SoilDataAvailabilityMap.pdf.

Courtesy of USDA-NRCS.

Speedy up-to-date data delivery service

In August 2005, the NRCS launched WSS. The site helps the USDA meet two of the federal budget management agenda items: eliminating paperwork and making government services accessible to more people. Before WSS went live on the Internet, soil survey maps and data were printed and bound into soil survey books that were free to the public at local USDA service centers, NRCS field offices, and public libraries. People could obtain the county's entire soil survey book and page through it to get specific information. If they needed information on parcels in multiple locations, they might have needed a separate publication for each.

Now, soil survey users can connect to the Internet, select a land parcel anywhere in the country, access the survey database, see the soil map of the specified parcel, generate soil interpretations, and download or print a PDF file. WSS saves a lot of USDA staff time. In the past, county office staff might spend an hour or two working with individual clients. Now, customers see the WSS site as self-service technology that they access at their business, residence, or public kiosk. This allows USDA personnel to allocate customer service time more efficiently.

In addition, the WSS has enabled NRCS to reach significantly more customers than it could reach with hard-copy publications. In the first six months of its existence, more than 200,000 users visited the WSS site. That number is nearly equal to the total number of soil survey publications that were printed in the previous six years combined. Electronic delivery also provides quicker access to up-to-date information.

WSS was created by the NRCS West Texas Telecommunications Project, Soil Survey Division, and Information Technology Center teams. They began the project by developing, in coordination with federal, state, and local representatives, a list of business requirements. The main goal was to meet the diverse needs of various users. NRCS also wanted WSS to become the official delivery mechanism for soil survey data, replacing the hard-copy publication process.

The WSS application suite was built using ArcIMS, ArcGIS Server, and ArcSDE software. Some server software is being fully integrated with the soil survey's master national database, which contains both spatial and tabular data. ArcIMS is the front end for delivering the application, and ArcGIS Server is the system delivering the information. In addition, ArcWeb Services provides address lookup functionality. The user types in an address, and ArcWeb Services returns the map view of that area.

The front-end application accesses the NRCS Soil Data Mart, which is the centralized repository and master database of the survey's soil spatial and tabular data that keys into the Web site. The ArcView application Soil Data Viewer renders soil interpretations on colored thematic maps. On the interactive GIS map, users can outline an area of interest (currently limited to a maximum of 10,000 acres) and select a usage option. The application provides suitability ratings of the area for the selected land use. WSS provides a single access point where users can retrieve current soil survey data as well as archived publications available in PDF for viewing and downloading.

The Soil Data Mart ensures that users get a consistent view of the data whether that data is accessed from a county office, state office, or private residence. The Soil Data Mart contains the current version of official soil data (spatial and tabular) for each soil survey area that is accessed by WSS. A transactional database is routinely maintained by soil scientists throughout the United States, who keep up-to-date data for every survey area in the nation. Periodically, a snapshot of the transactional database is certified and uploaded to the data mart. Earlier dataset versions are maintained in the Soil Data Warehouse and made available for other analyses such as temporal comparisons. The NRCS also employs various Web services for access to plant information, ecological site data, and local county offices' maps.

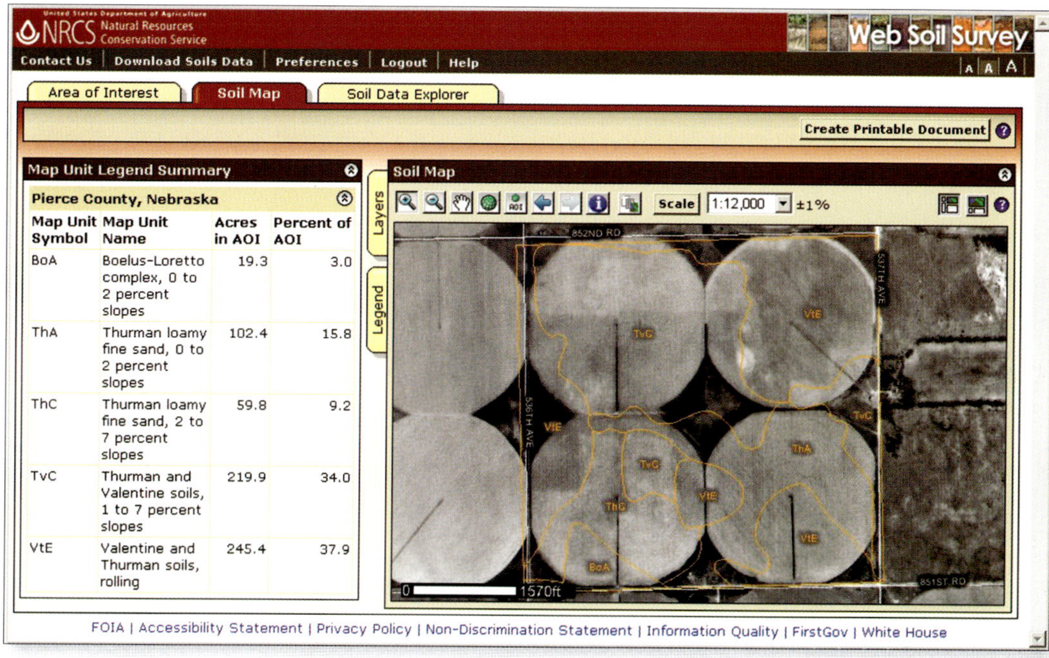

Figure 2.2 Web Soil Survey provides users with the ability to select a location anywhere in the United States and access soil data. Here the user has selected an area of interest (AOI) and is viewing the soil map for the area. The table on the left lists the soil map units found in the AOI and the acreage of each unit.

Courtesy of USDA-NRCS.

In the first six months of its existence, more than 200,000 users visited the WSS site. That number is nearly equal to the total number of soil survey publications that were printed in the previous six years combined.

Accessing interpretations with ease

WSS offers approximately fifty national standard interpretations. Users can select objectives such as installing a septic tank or sewage lagoon, building a house, or determining the production potential for different types of crops. For example, a farmer considers purchasing a farm and wants to evaluate which farm will best suit his or her needs. Using WSS, the farmer locates each farm on an interactive map and selects an interpretation, such as the production potential for corn or cotton lint, and the application accesses the database soil types to determine which farm would be most productive for the farmer's needs. The user is able to view the criteria used to generate the interpretation, a breakdown of the different soil components within the interpretations, and opportunities and constraints. A color-coded spatial representation gives the user an idea of the selected interpretation's levels of suitability, such as good, fair, and poor.

Other interpretations are wastewater treatment site selection and land application methods; productivity of crop land, range land, and forest land; and suitability/limitations for different recreational developments such as paths and trails, campsites, and picnic grounds. Sanitary facility site interpretations include landfill sites and the material for covering the landfill. Urban interpretations help city and county officials manage development. A user considering building a home can find information that may lead to reinforcing the building's foundation. Other interpretations are useful for a number of other user needs including road construction.

Adding new functionalities to WSS is a constant endeavor. Dennis Willamson, Texas GIS specialist for NRCS explains, "Eventually, USDA customers will be able to log into their USDA case files and drop in information, such as property boundaries, for quick access to information about selected properties in their files."

WSS is the beginning of a new era for NRCS and the delivery of soil survey information. The feedback on the site has been extremely favorable. WSS administrators have already received a great number of compliments from Web site users. One enthusiastic Web site visitor exclaimed, "The Web Soil Survey may be the greatest program that the federal government has ever provided!"

GIS community mutualism pays off in a New Mexico fire emergency

C. Randall Mynard and Gordon N. Keating, Los Alamos National Laboratory; and Paul M. Rich, Creekside Center for Earth Observation (formerly GIS team leader in GISLab, Earth and Environmental Science, Los Alamos National Laboratory), www.lanl.gov

Many of us watched and listened with trepidation to news reports of the May 2000 Cerro Grande wildfire as it swept through Los Alamos, New Mexico, burning approximately 43,000 acres. Captivating as the reports were, the media did not fully portray the conditions under which many individuals valiantly worked behind the scenes to provide support to emergency responders.

GIS technology was an integral part of emergency response during the fire, and it continues to support postfire restoration and environmental monitoring. Under often dangerous circumstances and for long hours, GIS staff from Los Alamos National Laboratory (LANL) and volunteers from other organizations worked to produce maps and provide support for emergency managers. After the fire, LANL GIS groups supported the multiagency (U.S. Postal Service, U.S. Forest Service [USFS], Bureau of Indian Affairs, U.S. Department of Energy [DOE], LANL, County of Los Alamos, Pueblo of Santa Clara, and Pueblo of San Ildefonso) Burned Area Emergency Rehabilitation (BAER) team with GIS data and maps for planning mitigation efforts. These GIS efforts during and after the fire offered important lessons about institutional matters, working relationships, and emergency preparedness.

The setting

Los Alamos is a small city of 11,000 people in north-central New Mexico, located about twenty-five miles southwest of Santa Fe at the foot of the Jemez Mountains. Home of the world-famous DOE Los Alamos National Laboratory, the town nestles amid a dense forest of pine, spruce, fir, and aspen that drapes the steep mountain slopes. The irregular surrounding terrain impedes highway access, and the heavily forested canyons facilitate the rapid spread of fire. Both are important factors in combating forest fires that sometimes threaten the area.

LANL stretches between Los Alamos and White Rock, a suburb to the southeast, and consists of nearly forty-five square miles of forested canyons and mesas. Isolated clusters of experimental facilities, testing areas, and research laboratories are scattered across this

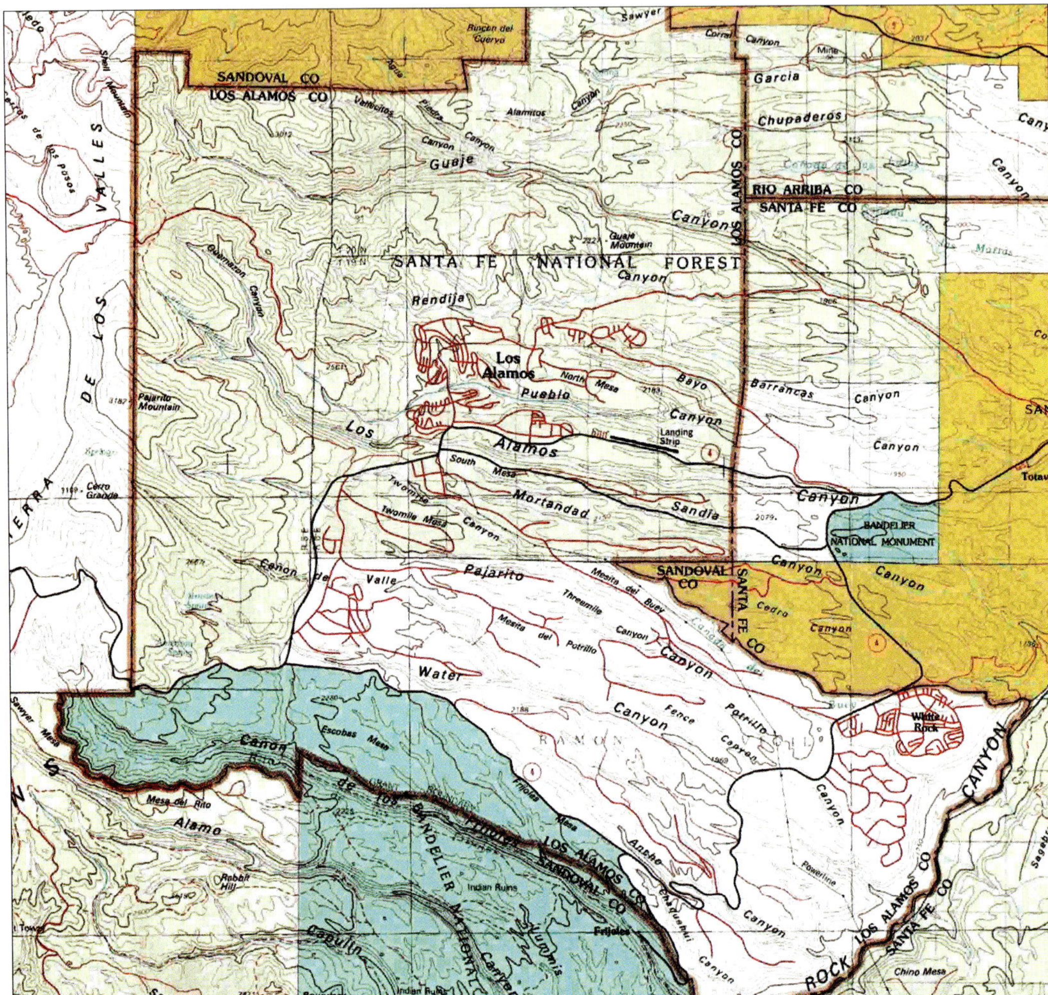

Figure 2.3 Topographic map of the Los Alamos area. Cerro Grande, where the fire began, is on the left.

Courtesy of U.S. Geological Survey.

landscape. The laboratory, operated by the University of California (Los Alamos National Security, LLC, as of 2006) for DOE, is best known for its weapons research but is also considered one of the nation's outstanding science and technology laboratories in many other fields including physics, climate change, computer science, biology, ecology, and geology.

LANL research has long involved high explosives and radioactive and hazardous materials. Certain locations at the laboratory, called potential release sites (PRSs), are suspected of having contamination from past activities, and they are regularly studied to determine if environmental remediation is necessary. PRSs could be a concern if fire or soil erosion threatens to disperse contaminants found to be present.

Facing an emergency

On May 4, 2000, National Park Service (NPS) personnel ignited a prescribed fire on Cerro Grande, a mountain four miles southwest of Los Alamos. Prefire conditions seemed appropriate for a controlled burn, but by the next day, the fire had escaped its containment and began advancing eastward, pushed by strong winds. The blaze, then considered a wildfire, began spreading toward LANL, entering LANL property and the town of Los Alamos by May 10 and causing significant damage.

By May 18, the fire was only 80 percent contained. Los Alamos and White Rock had been completely evacuated, and the laboratory remained closed until May 22, a shutdown unprecedented in its 57-year history. While the fire took no human lives, 429 families lost their homes, and many structures were burned, including 112 small structures at LANL.

Figure 2.4
Smoke from the Cerro Grande fire could be seen from miles away.

Courtesy of U.S. Forest Service.

The public and the local government were concerned about chemical or radiological contaminants that might be present in smoke or later be washed from denuded soils. LANL emergency managers also needed as much information as possible about the fire, including where the fire was actively burning and where it had burned, what facilities were at risk, what structures had been burned, what infrastructure and utility damage had been sustained, where new fire roads were located, and which PRSs were burned or might be in the fire's path. Analytical data about possible contaminants in the affected PRSs were also needed.

Communicating a need

Despite the evacuation, the need for GIS support was evident, and the challenge was how to provide it from an off-site location and how to access the necessary data, which was diverse, formatted in various ways, and located on numerous servers at LANL and elsewhere.

As the fire raged, USFS had a GIS capability for preparing daily operations maps for fire crews and for briefing the news media, but specialized maps of LANL facilities were also needed. Sandia National Laboratories, a DOE research and development laboratory in Albuquerque, began supporting the emergency activities with fire maps that were used to make essential decisions about LANL assets that might be at risk. The Environmental GIS office at Sandia served primarily as a data gathering and distribution node and provided fire perimeter data needed by DOE teams who were monitoring radiation and chemical releases in and around LANL.

On May 13, Denise Bleakly of the GIS office at Sandia began gathering data by first requesting assistance from members of the New Mexico Geographic Information Council. At no cost, she quickly received regional GIS datasets from the Bureau of Land Management (BLM), USFS, and the University of New Mexico Earth Data Analysis Center. This data enabled Sandia to prepare initial fire maps.

Meanwhile, LANL GIS staff had begun to assemble a team in Santa Fe to assess the impact of the fire on the PRSs. One of the main GIS facilities at the laboratory, the Facility for Information Management, Analysis, and Display (FIMAD), was the primary data repository and GIS resource for the lab's large Environmental Restoration (ER) project that produced more than a thousand new maps each year. During the fire, FIMAD was closed due to the lab's shutdown, but its computer servers continued to function unattended. The servers were connected to the lab's Intranet, which also remained online. That this link to the FIMAD servers remained was critical to off-site GIS efforts.

Two LANL staff members who lived outside the evacuation zone and were working at home on assignments, combined the data they had on hand, which was sufficient for creating basemaps of the laboratory. For a GIS facility, the ER project made use of an existing subcontract with the Loft4 Digital Media Studio in Santa Fe. To support FIMAD efforts during the emergency, a safe place was needed to store sensitive project data, and Sandia offered this service.

Within forty-eight hours, the linkage between mission concept, basic equipment, GIS expertise, data for maps, networking experience, and office space was achieved, and the group was able to establish an improvised GIS facility at Loft4 using the additional staff and equipment there.

GIS for emergency response

Managing the emergency response to the Cerro Grande fire was made difficult and complex by the fire's rapid progression, its changing location, and the fact that the fire divided into two fronts. Constantly updated maps were critical for LANL managers and firefighters to decide which response to make and how to deploy the hundreds of personnel on the fire lines.

BLM conducted nightly overflights to collect location data of the rapidly moving fire. BLM staff would digitize the infrared imagery, create polygon or line files, and relay this data each morning to the GIS facility at Sandia, where the data was converted to spatially referenced overlays of fire perimeters and hot spots overlaid onto basemaps. Sandia electronically transferred these files to the FIMAD server at Los Alamos or directly to Loft4. LANL staff at Loft4 used them to create hundreds of maps that showed the expanding burned areas in relation to contaminated areas (PRSs). The GIS efforts continued twenty-four hours a day for ten days. LANL and Loft4 personnel, divided into shifts, quickly developed team roles and responsibilities, which proved effective during the demanding hours of around-the-clock operations.

The LANL Emergency Operations Center staff used the maps produced by FIMAD staff in Santa Fe to plan the laboratory's response to the changing fire threat, mark locations

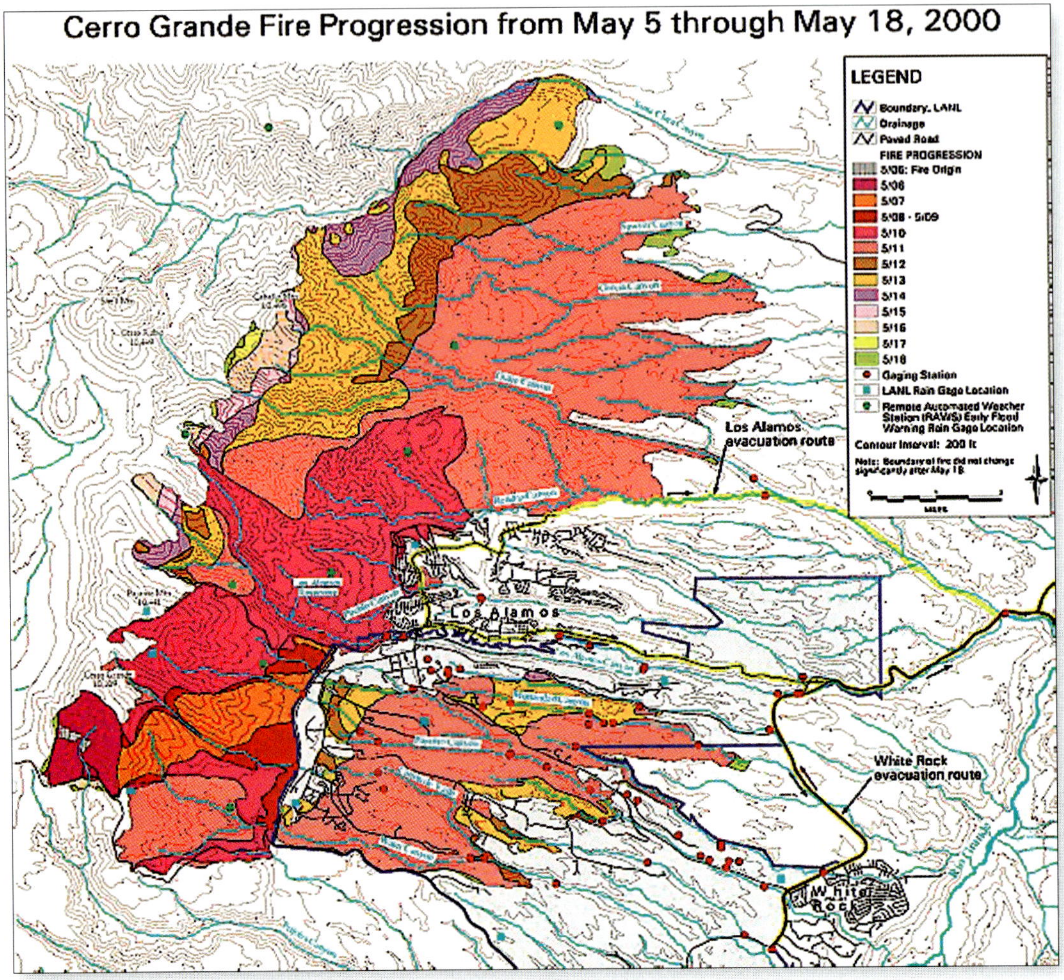

Figure 2.5 Cerro Grande fire progression map showing areas burned on successive days.

Courtesy of GISLab, Los Alamos National Laboratory.

A coordinated effort from the GIS community is necessary for cross-institutional integration of GIS technology. The fire illustrated how important it is for staff to work together and how effective GIS can be when cooperation is optimal.

of new fire roads, and identify at-risk, damaged, or destroyed structures. Using a range of runoff and erosion computer models, LANL staff was able to perform several modeling runs while the fire was still burning strongly. The models were used to identify areas that posed threats to humans, infrastructure, and the environment should a large rain event occur.

Postfire collaborations and decision support

The BAER team arrived on Monday, May 15, and began a preliminary assessment of the fire area. BAER teams are formed after major fires to assess damage caused by the fire and to implement a rehabilitation plan that would prevent loss of life and property and reduce further natural resource damage. BAER teams are composed of highly skilled wildlife biologists, archaeologists, soil scientists, landscape architects, geologists, ecologists, engineers, foresters, botanists, GIS and GPS specialists, and other professionals from around the nation. Team members included personnel from the USFS, BLM, NPS, USGS (United States Geological Survey), and LANL.

In addition, a multiagency team was formed to begin emergency rehabilitation for the Cerro Grande fire. The Multiagency Coordination (MAC) Team, formed on May 15, included representatives from all the landowner agencies and acted as the umbrella organization during the assessment and rehabilitation. The goal of MAC Team is to provide interagency communications and minimize red tape.

As the BAER team and LANL began to assess damage and plan recovery efforts, the need for infrastructure information and maps of fire severity increased dramatically. Fortunately, none of the LANL GIS facilities was damaged or destroyed in the fire, and FIMAD and other LANL GIS offices quickly responded with coordination, computers, and logistical support and provided data and supplemented maps that the BAER team produced.

The BAER team arrived while the fire was still burning, and LANL GIS staff worked closely with the team to determine burn severity patterns in the laboratory's forests and grasslands and to provide information about ecological, biological, and archeological resources at LANL. Additionally, they provided the USFS Incident Management team with data on area fuel loading and forest structure for use in the service's fire behavior model, which helped to determine where to position firefighting assets.

On June 2, the Emergency Rehabilitation Team (ERT) formed to conduct initial assessments of potential environmental impacts from the fire. The team consisted of representatives from LANL groups and other organizations from outside the laboratory.

GIS analyses and modeling were important tools for the assessment team in determining the immediate and long-term threat of flash flooding and soil erosion. The results of these studies contributed to management decisions to close a facility found to be at high risk from flash floods, construct flood retention structures, and make improvements to bridges, culverts, and stream channels on LANL grounds and within Los Alamos County.

Because in some places the soil had become hydrophobic (water-repellent), flash flooding was a big concern. GIS analyses revealed the potential for stormwater discharge rates of two

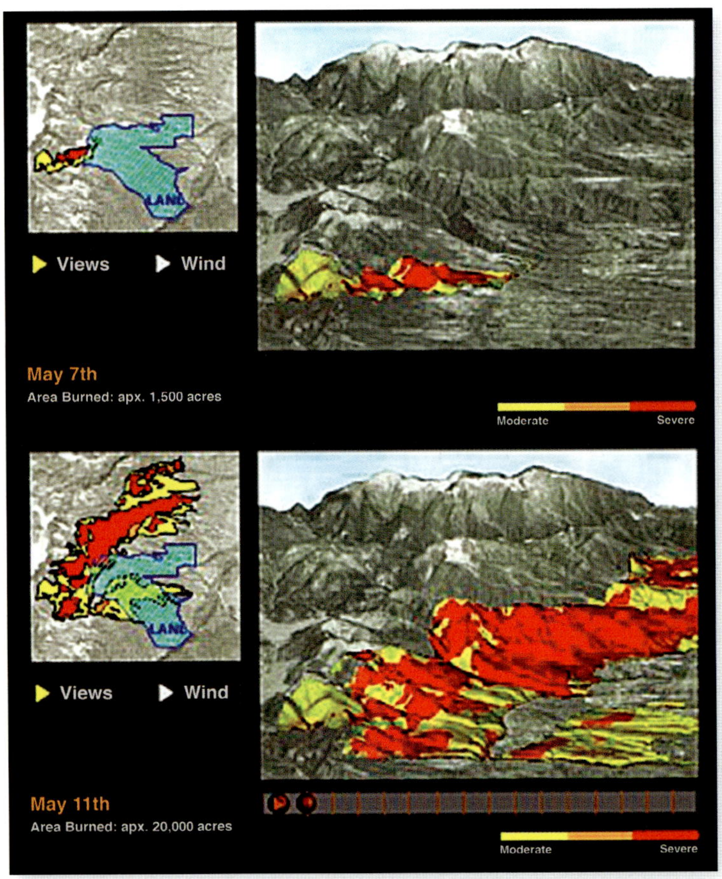

Figure 2.6 Cerro Grande Rehabilitation Project GIS Web pages with animated 3D views of fire progression and wind data.

Courtesy of GISLab, Los Alamos National Laboratory.

and a half to nearly six times normal and warned of the possibility of huge flash floods that could reach White Rock. This prompted the construction of a major retention structure. Additionally, FIMAD and the ER project staff provided data and maps about PRSs, particularly those located in burned areas, those affected by the fire, or those in drainage areas subject to erosion. Several hundred maps were produced in a few weeks and were used for geomorphological, geological, and sediment transport and migration studies. There were also several remote-sensing techniques used to determine the impact of fire on vegetation, erosion, forest fuel loading, and soils, and the GIS team used data from them for modeling future fire potential, forest regrowth, and runoff.

GIS support for rehabilitation

Soon after the fire, the Cerro Grande Rehabilitation Project (CGRP) was initiated to coordinate efforts to restore damaged infrastructure and property at the laboratory and to mitigate hazards resulting from the fire such as flooding, erosion, and dead and dying trees. These ongoing remediation and mitigation efforts increased the need for GIS at LANL.

In August 2000, CGRP funded a new GIS for its subprojects and as an institutional resource for research and operations. GISLab, the successor to FIMAD, manages these programs and several other tasks including planning for forest thinning, delineation of floodplain changes, input for upgrading GIS capabilities at the new emergency operations center, mapping infrastructure, and predictive fire modeling. These activities have improved LANL disaster response capabilities and have enabled a significant reduction in risk from future fires, floods, and other emergencies.

The GISLab spatial data warehouse is a major component of an enterprise GIS design that features large data storage capacity, geospatial search functions, and Web-based access to both spatial data and metadata.

At its annual User Conference in July 2002, ESRI presented GISLab of LANL and the BAER team with its Special Achievement in GIS award, recognizing their GIS efforts during the fire and afterward in support of CGRP. Dr. Paul Rich, then team leader for GISLab said, "The events of May 2000 were extraordinary. The response of the GIS community was equally extraordinary . . . the most important preparation was in place—a dedicated community of GIS professionals skilled at solving problems."

Lessons learned and future plans

The Cerro Grande Fire taught decision makers at LANL many lessons about the use of GIS in emergencies, some that were immediately apparent and others that became evident later on, including:

- GIS provides an integrating framework for assessing natural and human hazards in a landscape context
- A strong GIS capability is needed for emergency response
- Coordinated emergency plans are needed for GIS operations
- A method for locating employees should be in place
- GIS data should be complete, backed up, and available during an emergency
- GIS procedures should be flexible so they can be adapted to the circumstances
- GIS resources should be fully integrated within the organization

A coordinated effort from the GIS community is necessary for cross-institutional integration of GIS technology. The fire illustrated how important it is for staff to work together and how effective GIS can be when cooperation is optimal. The efficiency and effectiveness of GIS efforts were dependent on interactions within the organization and with external agencies.

The Cerro Grande Fire was a serious regional emergency that provided neighboring governments and LANL with important lessons about how to deal with large-scale disasters. These lessons are being incorporated into local, state, federal, and LANL disaster planning so that future emergencies will be handled with improved procedures, communication, and coordination. Recently, efforts from the DOE GIS user group and other stakeholders have highlighted the benefits of coordinating GIS efforts across the DOE complex.

Dr. Rich has been representing LANL on the DOE Geospatial Science Steering Committee and helped lead an effort toward the recent formation of a new DOE Geospatial Science Program. This new program was established in 2005 to coordinate and maximize the DOE investment in geospatial activities as they support the DOE core mission.

References

Mynard, C. R., G. N. Keating, and P. M. Rich. 2005. GIS emergency support for the May 2000 Cerro Grande Wildfire, Los Alamos, New Mexico. *Journal of Emergency Management* 3:19–28.

Mynard, C. R., G. N. Keating, P. M. Rich, and D. Bleakly. 2003. GIS emergency support for the May 2000 Cerro Grande Wildfire, Los Alamos, New Mexico. LA-14007-MS.

All agencies need to plan for emergencies. Collaborative planning is a way to proactively address threats from wildland fires or other issues of regional importance in a way that brings multiple perspectives and diverse values into play in a harmonized fashion.

During the past two decades GIS has become an increasingly important and pervasive tool at all levels of government and in private industry. The events of September 11, 2001, and the conflicts in Afghanistan and Iraq have brought widespread public recognition of GIS as a key decision support and information-sharing tool for emergency management, intelligence gathering and analysis, and military operations. An estimated 80 to 90 percent of all government data has a geographic component. This is certainly the case at DOE, where geospatial science and GIS are used across the DOE complex to address a broad set of security, safety, cost-saving, and research and development matters. These include emergency response, homeland security, environmental restoration, weapons stewardship, site operations, transportation routing (nuclear materials, weapons, and hazardous waste), infrastructure siting and operations, and basic and applied scientific research.

TOXMAP: A GIS-based gateway to environmental health resources

Colette Hochstein and Marti Szczu, National Library of Medicine; Darren Gemoets, Aquilent, Inc., http://toxmap.nlm.nih.gov

In 1966, President Johnson's Science Advisory Committee issued a report entitled "Handling of Toxicological Information," which stated, "there exists an urgent need for a . . . more coordinated and more complete computer-based file of toxicology information than any currently available and . . . that access to this file must be more generally available to all those . . . needing such information."

This resulted in legislation authorizing the U.S. Department of Health and Human Services to establish a Toxicology Information Program. The program was formed in the late 1960s at the National Library of Medicine (NLM), which is part of the National Institutes of Health. One of the NLM strategic goals is to facilitate and encourage access to and use of high-quality information. It was later expanded to include environmental health and was renamed Toxicology and Environmental Health Information Program (TEHIP). Over the years, TEHIP (http://sis.nlm.nih.gov/enviro.html) has developed a collection of databases and related information on environmental health and toxicology. Some of these resources were created at NLM; others originated at a variety of national and international agencies and associations.

In 1986, Congress enacted a law called the Emergency Planning and Community Right to Know Act. This law was designed to help local communities protect public health, safety, and the environment from chemical hazards. The law requires certain U.S. facilities to report annually on their releases of specified chemicals; it pertains to industries with ten or more full-time employees that manufacture, process, or use more than 25,000 pounds of any one specified Toxics Release Inventory (TRI) chemical. The reports contain information about the types and amounts of chemicals released each year into the air, water, and ground and by underground injection. There is also information about their transfer to waste sites and waste treatment methods and efficiency.

Before the Internet, NLM collaborated with the U.S. Environmental Protection Agency (EPA) to make TRI accessible to online subscribers. The EPA now offers comprehensive information about the TRI program on its Web site (www.epa.gov/tri/). A 2003 EPA report discusses the various ways in which the TRI data is being used by citizens, government, business, and academia. As a service to its users, NLM has also continued to provide TRI data to its community through a multidatabase Web application called TOXNET (http://toxnet.nlm.nih.gov). TOXNET contains an extensive collection of bibliographic and research information in the areas of toxicology and environmental health, as well as data excerpted from the literature on hazardous chemical substances. TRI contains discrete geospatial data fields such as latitude/longitude, street address, city/state, and ZIP Code in each record. Searches on location parameters such as city, state, and ZIP Code return a summary list of matching facilities with information about each of the chemical substances for which the facility has submitted a report.

TOXMAP inception

The result is a resource that can show the geographic distribution of releases by chemical and can link to information in the TOXNET Hazardous Substances Data Bank and other authoritative resources. It uses chemical and geographic terms from displayed maps to search bibliographic databases and integrate geographically coded data from other sources. Easy to navigate and understand, it is accessible by a broad range of users.

Work on developing the TOXMAP information system began in 2001, and a beta version was launched in January 2003. After an interface redesign in June 2004, TOXMAP was released to the public in September 2004. Its first major revision occurred in early 2006.

TOXMAP has one project lead and two developers and uses ESRI ArcIMS 9.1, which supports the needed functionality and customization. The ESRI Place Finder Web service enables searching for cities, ZIP Codes, counties, and other places in the United States. TOXMAP hardware consists of a Dell PowerEdge 2650 server with dual 3.0 GHz Intel Xeon processors.

TOXMAP also uses open-source tools. Written in Java, TOXMAP uses software from the Apache Software Foundation, including Apache HTTP Server, Tomcat, Struts, JMeter, and Ant. It also uses the open-source relational database application MySQL, which has provided cost-savings as well as scalability and speed. A nationwide basemap was developed from data available from the U.S. Geological Survey National Atlas.

Creating maps with TOXMAP

TOXMAP helps users create national or local maps that show sites where TRI chemicals are released into the air, water, and ground. It identifies the releasing facilities, color codes release amounts for single years, and provides multiyear chemical-release trends, starting with 1987. Users can search the system by chemical name, name fragment, and location such as city, state, or ZIP Code.

TOXMAP also includes Superfund program sites, which are sites throughout the United States and its territories that contain substances either designated as hazardous under the Comprehensive Environmental Response, Compensation, and Liability Act (CERCLA) or identified as such under other laws. CERCLA designates more than

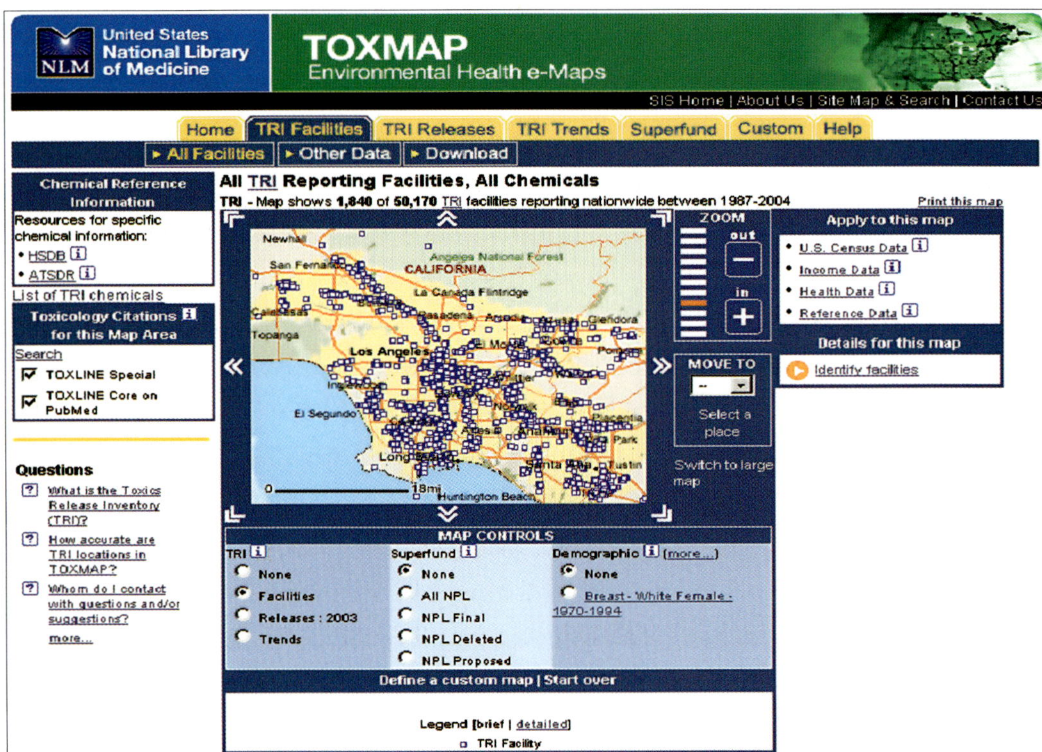

Figure 2.7 TRI facilities in the Los Angeles area.

Courtesy of National Library of Medicine.

eight hundred substances as hazardous and many more as potentially hazardous to human health or to the environment when improperly treated, stored, transported, disposed of, or otherwise managed. Sites where releases or potential releases have been reported are listed in a searchable EPA database called the Comprehensive Environmental Response, Compensation, and Liability Information System (CERCLIS).

TOXMAP provides users with links to research studies and other related health information about the Superfund contaminants. The National Priorities List (NPL) contains the most serious uncontrolled or abandoned hazardous waste sites throughout the United States and its territories and serves as an information and management tool to guide the federal government in the site cleanup process.

There are several ways to start making maps with TOXMAP. Users can select the tabs at the top of the Web page, enter chemical or locational terms into the Quick Search box, or click on the map of the United States. TRI facilities are denoted by colored circles; each color represents the chemical release amount. Ranges are shown in the legend. Users can hover over each color with the mouse or click "detailed" to change the legend type. Standard map panning and zooming are provided.

Users can select Identify Releases under Details for this Map for a list of TRI releases, starting with the largest. A list of facilities appears to the right of the map, and details are provided below the list. The amounts of release into the air, water, and ground, and through underground injection, are displayed in the Emissions Estimates table under each facility name. The TRI Trends tab provides information on changes in TRI chemical releases over time.

The Chemical Reference Information section at the top left of the TOXMAP page provides related technical information, such as human health effects and probable routes for human exposure, from the NLM Hazardous Substances Data Bank. More information is provided via links to the Agency for Toxic Substances and Disease Registry.

The section Toxicology Citations for Chemical and Map Area can be used to search the NLM PubMed journal literature search system and the TOXLINE Special database for geographically related items, combining chemical terms with names of the cities, rivers, and lakes in the area.

By clicking on the Superfund tab the user can view Superfund sites on the National Priorities List, their status on the NPL, a list of contaminants at each site, and the site's EPA progress profile report.

Other data can be applied to the map, including mortality data from the Centers for Disease Control and Prevention and the National Cancer Institute's Division of Cancer Epidemiology and Genetics, income data from the U.S. Bureau of Economic Analysis, and demographic data from the 1990 and 2000 U.S. Censuses. TOXMAP also helps users create a custom-made map via the Define a Custom Map tab.

Because some users may be unfamiliar with GIS, TOXMAP provides helpful tools such as bar charts and frequently asked questions (FAQ). Bar charts can help provide accurate understanding of data, trends, values, and relationships. For example, users can select the TRI Facilities tab to see a bar chart for each facility, showing up to five of the top reported chemical emissions. Similarly, the TRI Releases tab provides bar charts showing distribution by medium (land, air, water, and underground injection) for the emissions of the selected chemical at each facility. The TRI Trends tab presents bar charts of emissions estimates for the most recent fifteen years for the selected chemical, as reported annually to EPA. The FAQ list under the Help tab can supplement the user's understanding of map displays and data and prevent misinterpretation.

TOXMAP can be used to generate specific custom-made maps from other Web sites via a URL, avoiding the need to conduct a search and promoting access and use by a wide audience. A single HTML tag instructs a Web page to go to TOXMAP. Information about the requested custom map is passed to TOXMAP through a URL that includes specific information about the type of map desired.

Traffic on a Web site can be one indicator of return on investment and, in this case, of improved access to information. Since its public release in September 2004, TOXMAP has received more than 100,000 visits generating over 650,000 maps, and more than six million total hits.

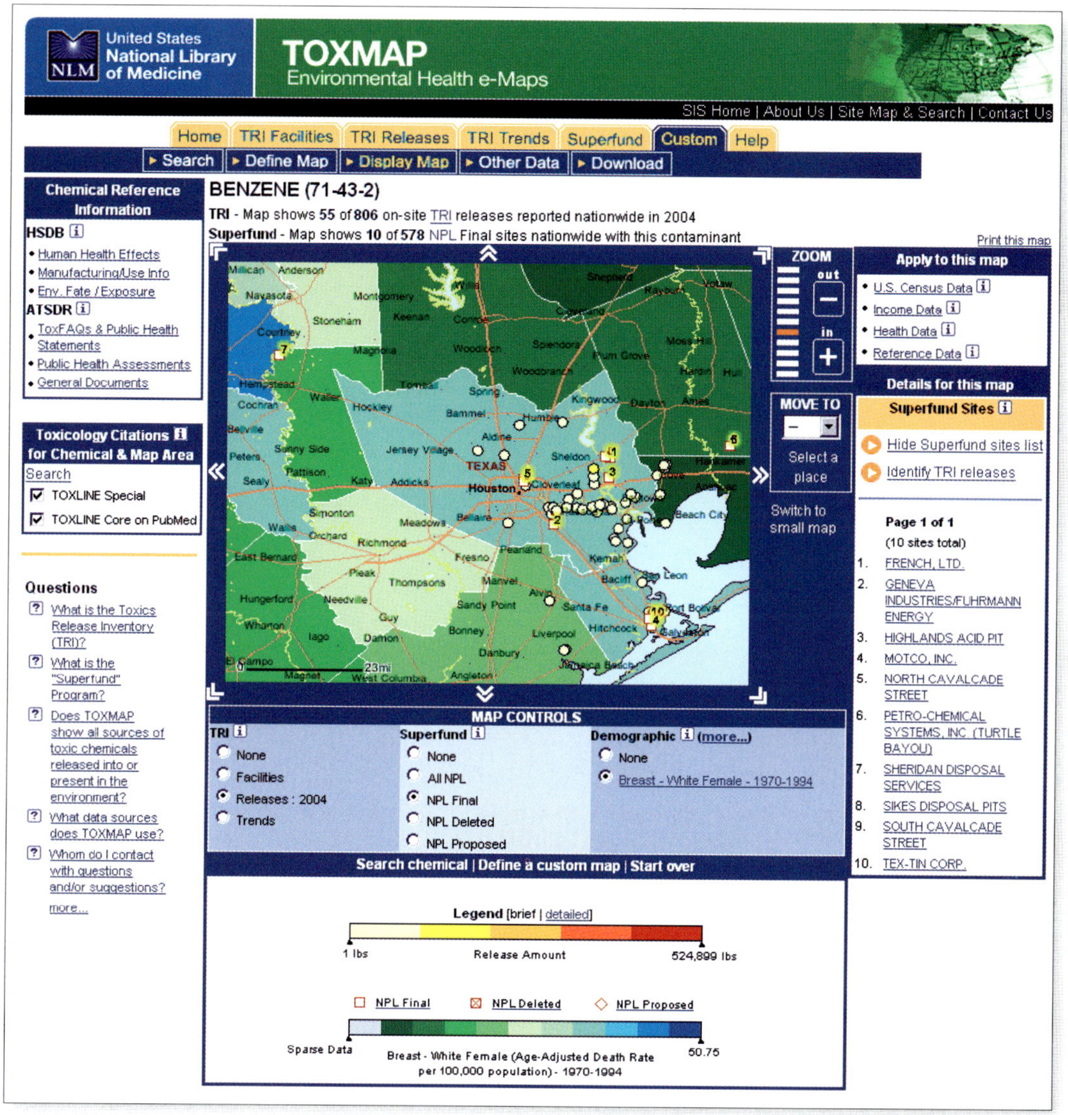

Figure 2.8 Benzene TRI releases, final NLP Superfund sites, and white female breast cancer mortality for 2004 in the Houston area.

Courtesy of National Library of Medicine.

User feedback

Although a GIS offers many benefits, some applications can be a challenge to navigate. Involving users in the development of a resource is crucial to maximizing its usefulness. To enhance the usability and effective presentation of TOXMAP, NLM has performed a variety of usability studies, including conducting intermittent, informal usability tests on various versions; requesting feedback from specific users such as health professionals; and reviewing feedback from user e-mails. The NLM also conducted two professionally moderated, Web-based focus groups on TOXMAP in 2005, which provided detailed usability comments.

Based largely on this feedback, NLM recently added Superfund sites, contaminants, and additional demographic layers (age, gender, income, cancer mortality, and disease mortality) to TOXMAP. Other features are planned, such as the ability to display multiple chemicals on a single map; updated and expanded U.S. Census demographics; and the ability to display one type or a combination of types of releases (air, water, land, and/or underground injection).

TOXMAP provides users with a new way to explore and understand a number of toxicology and environmental health resources. Its distinct approach has been especially useful for developing a resource that engages a technology (GIS) and a subject (hazardous chemicals) that may not be familiar to many users. The use of professionally moderated focus groups has provided cost-effective, detailed feedback from a broad spectrum of NLM users and has helped lay the foundation for future collaborations with both the NLM user community and professional organizations.

Traffic on a Web site can be one indicator of return on investment and, in this case, of improved access to information. Since its public release in September 2004, TOXMAP has received more than 100,000 visits generating over 650,000 maps, and more than six million total hits.

Mapping school district demographics on a national level

Tai Phan, mathematical statistician, National Center for Education Statistics, Institute of Education Sciences, U.S. Department of Education

The National Center for Education Statistics (NCES) collects statistics on education in the United States, analyzes and reports their significance, and assists states, local education agencies, and postsecondary institutions in improving their statistical systems. NCES conducts research and provides policy-relevant data on issues as diverse as enrollment trends, access of minorities to postsecondary education, academic achievement of students, comparisons of the U.S. education system with education systems in other countries, and the association between education and economic productivity.

In the early 2000s, NCES, the principal statistical agency within the U.S. Department of Education, implemented a GIS using ESRI ArcIMS and ArcSDE to disseminate decennial census and administrative data about public schools and school districts. The public Web site (http://nces.ed.gov/surveys/sdds) provides demographic and geographic information about local education agencies in the fifty states and the District of Columbia.

A national mapping Web site

The mapping of education data can make clear where issues are geospatially linked and facilitate a variety of practical actions, the development of effective policy, and the emergence of research insights. NCES has long identified and collected data on public schools and school districts. However, although localities have characteristically known the boundaries of their school districts and depicted them on paper maps, school district boundary information had been unavailable nationally for the purposes of dynamic electronic mapping. It was not until early 1990 that NCES, in coordination with the U.S. Census Bureau, was able to define the geospatial boundaries of districts and map them electronically.

The NCES mapping site has evolved since its inception in 2002 to include a broad range of school district data. Specific areas can be displayed by pointing and clicking on a map or by searching for a particular locality. The NCES mapping system is the only system in the nation to visually link educational organizations (in other words, schools within

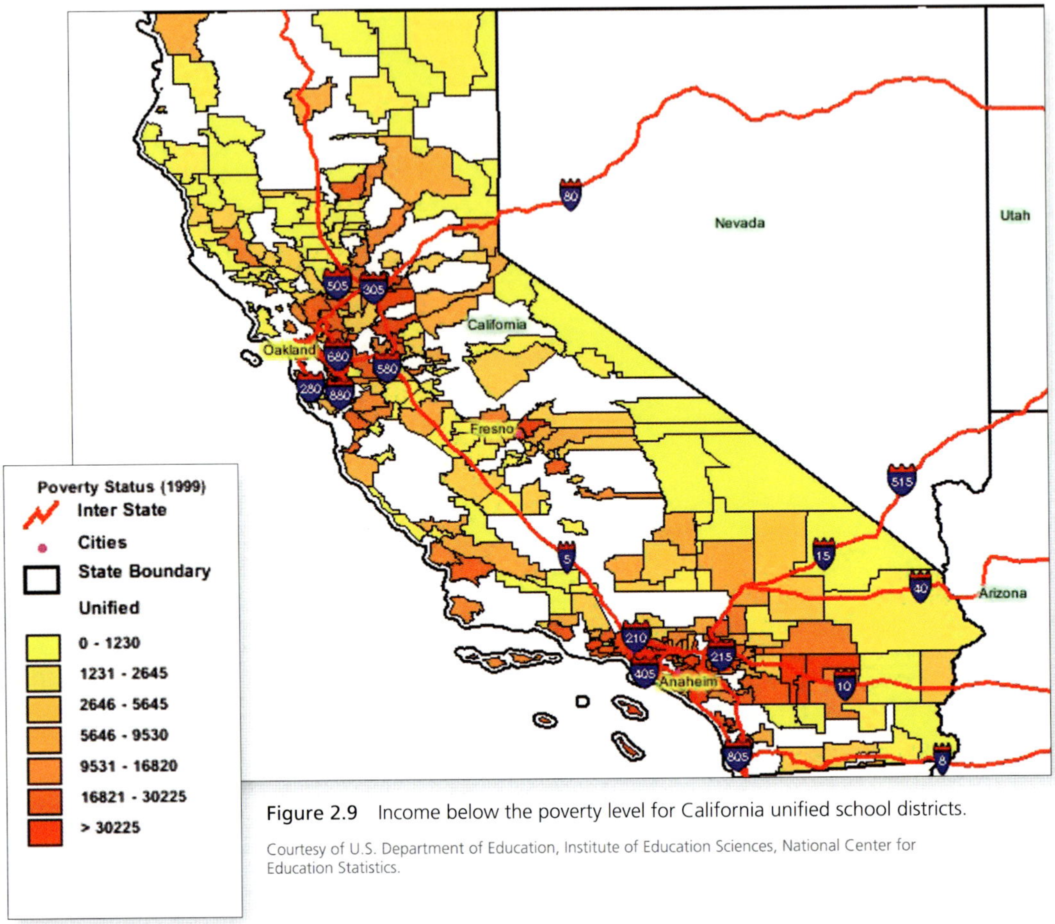

Figure 2.9 Income below the poverty level for California unified school districts.

Courtesy of U.S. Department of Education, Institute of Education Sciences, National Center for Education Statistics.

school districts and school districts within states) to their geographical locations and to demographic and economic information about the respective populations.

NCES has mapped school district boundaries through the Title I program (Improving the Academic Achievement of the Disadvantaged) and developed relevant layers of information based on two primary sources. One source is census data that characterizes the population according to race/ethnicity, age/gender, employment/income, and household composition. The other source is Common Core of Data (CCD), the annual NCES collection of data about schools and school districts that can be used to identify trends in enrollment, completion, student characteristics, and location.

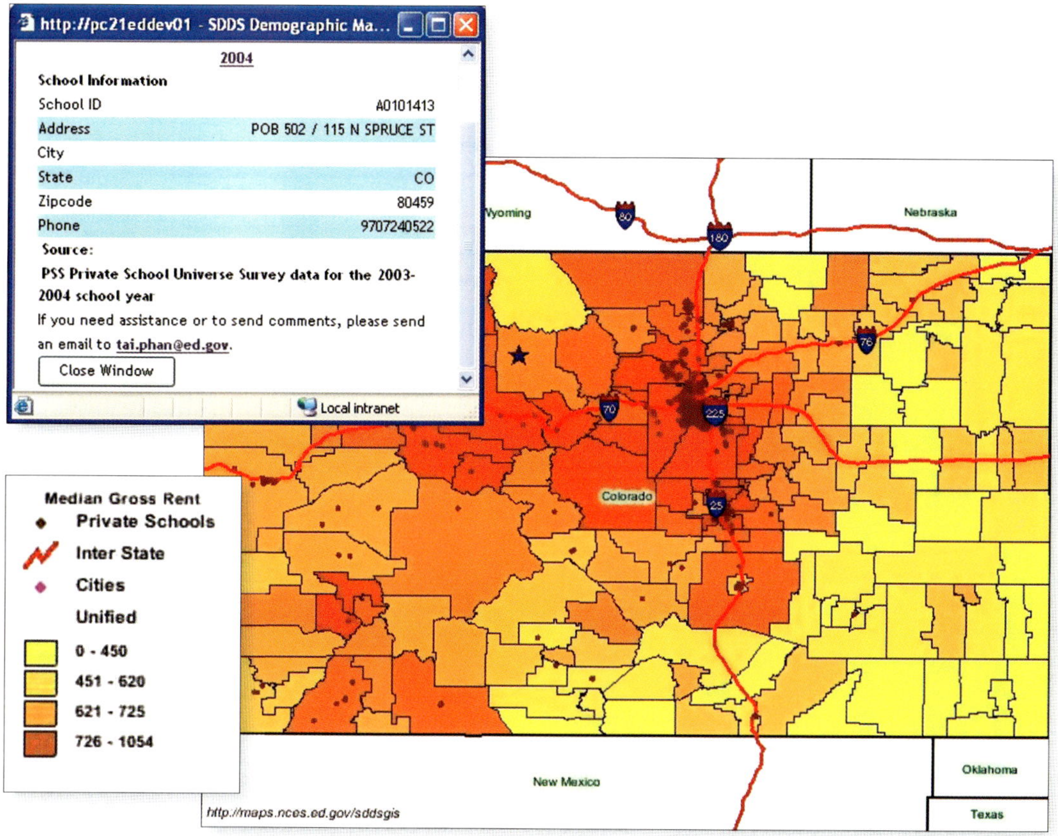

Figure 2.10 Median gross rent and locations of private schools for the state of Colorado.

Courtesy of U.S. Department of Education, Institute of Education Sciences, National Center for Education Statistics.

GIS data from the NCES Web site can help with the planning and funding of new schools, classrooms, bus routes, courses, textbooks, teachers, and library space. Effective mapping can help determine the most appropriate distribution of funding throughout a state or a school district.

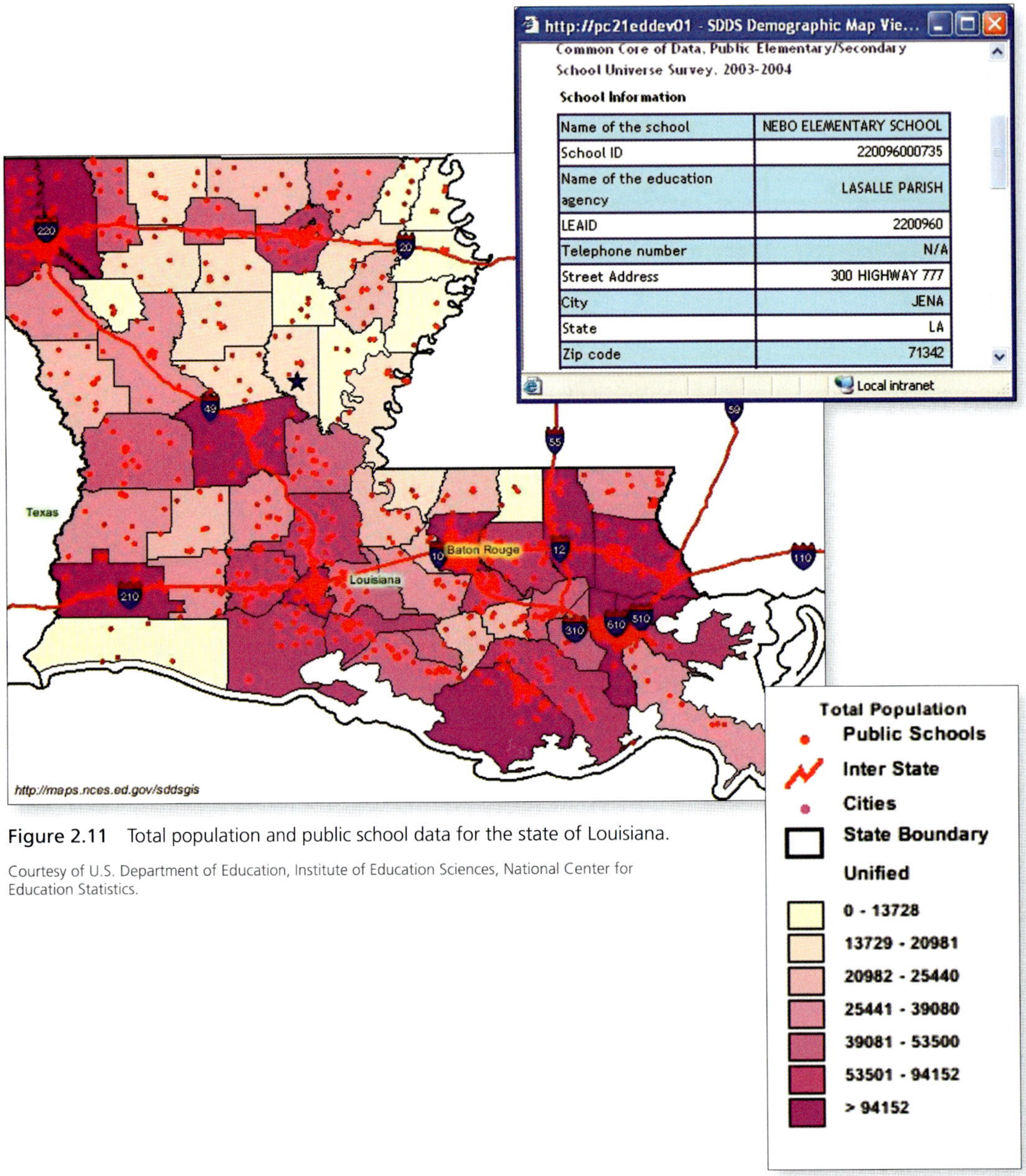

Figure 2.11 Total population and public school data for the state of Louisiana.

Courtesy of U.S. Department of Education, Institute of Education Sciences, National Center for Education Statistics.

Each datum from the census is linked to the boundary layer of a state, county, or school district. This enables viewing population ecology as related to the state, county, or school district and to the school or school district student membership (including student attributes such as race/ethnicity, grade, and gender). Multiple years of school and district data from the CCD are also linked to each geospatial entity relevant to education.

NCES maps in action

The NCES mapping Web site can help local, state, and national policy makers stay on top of changing population trends within specific areas and anticipate future needs for resources.

Disaster prevention and recovery can be more effectively managed with access to detailed information about the population of the area in question. For example, in 2005, Hurricane Katrina affected a large area that included many schools. Awareness of the number of students in the affected area and their geographic distribution could have facilitated preparation and recovery.

Social and economic problems can be addressed with effective allocation of teachers, transportation, and buildings, based on knowledge about changes in the population and the capacity of the education system. GIS data from the NCES Web site can help with the planning and funding of new schools, classrooms, bus routes, courses, textbooks, teachers, and library space. Effective mapping can help determine the most appropriate distribution of funding throughout a state or a school district.

One of the uses of GIS in data analysis facilitates spatial visualization of student demographics and school district attributes as well as their relationships to schools, parents, and counties. This helps researchers examine whether parents' income or characteristics such as parents' occupations or number of wage earners can be used to measure the children's likelihood of success in school. The mapping of changing school district demographics can help researchers pinpoint locational factors and the associated configuration of socioeconomic attributes within spatial boundaries. These observations and insights can help guide educational policies and effective allocation of educational resources.

3

Making Informed Decisions

Federal agencies are faced with the challenge of making choices that simultaneously serve their mission, adhere to policies and laws, and avoid or mitigate negative effects. Agency analysts need to be able to create multiple scenarios based on the best information available, propose and evaluate alternatives, and measure the impacts. GIS is the optimal decision support tool because it enables staff to assemble and process data from disparate sources, illustrate workflows, model dynamic natural and man-made events, and easily recognize changing situations by visualizing the long-term effects of actions.

In a world where information abounds, the art of implementing effective public policy has become more challenging as decision makers try to assimilate an endless stream of facts. To make informed choices, decision makers need to be aware of all issues that affect a proposed activity or project. They need access to departmental and organization-wide data repositories and decision support tools to help them understand the data. Rather than being bogged down with deciphering numerous and varied reports derived from multiple departmental databases, agency decision makers gain a definite business advantage from a GIS that integrates data into one easily interpreted source: a map.

Increasingly, organizations are using GIS technology—particularly three-dimensional visualization—to view what-if scenarios. GIS software makes it possible for analysts to model various conditions by adding, modifying, or taking away components or changing the values of variables. An analyst can diagram a real or theoretical process, such as viewing areas that would be affected by a five-year or one-hundred-year flood, and make predictions or calculations about the results. Analysts and planners can generate and view plausible alternatives, recommend a plan of action, articulate the issues more clearly, or monitor a project's progress.

Nearly every public policy issue has a geographic component. Knowing the *where* is paramount to the other considerations of accountability. GIS technology is becoming an integral part of the decision-making process as it helps to shape and influence choices.

Using GIS to protect drinking water

William Cooter, Jay Rineer, Danette Boezio, and Sunil Rao, RTI International; and A. Roger Anzzolin, U.S. Environmental Protection Agency, www.epa.gov/safewater

In 2002, the U.S. Environmental Protection Agency (EPA) Office of Ground Water and Drinking Water (OGWDW) identified a need to provide key EPA technical staff and managers with a readily accessible tool to run queries and reports and generate maps vital to the efficient management of drinking water programs under the Safe Drinking Water Act (SDWA). The Drinking Water Mapping Application (DWMA) is a Web-based GIS tool that serves this purpose.

Since September 11, 2001, maintaining controlled access to critical information has become a priority. Ensuring security with regard to accessing information that can reveal both locational information on drinking water intakes or wellheads and GIS data for managing risks to public drinking water supplies has become even more important. Deployed on the EPA Intranet in July 2003, DWMA is the first GIS application and analytical tool developed by EPA that focuses on public water supplies, source water protection, and underground injection control (UIC). OGWDW and other users at EPA headquarters and regional offices have benefited from improved access to information, cost and time savings, and enhanced technical analyses provided by DWMA.

DWMA is part of the EPA Watershed Assessment, Tracking, and Environmental Results architecture. For surface water, DWMA uses the U.S. Geological Survey (USGS) national hydrography dataset (NHD) to create an analysis framework, and within the existing EPA database infrastructure it connects directly to the Reach Address Database (a repository of water quality data submitted by states to EPA) and other EPA data systems. DWMA provides a hydrogeologic framework for classifying groundwater environments and offers both land surface and geological cross-section displays for areas of investigation around UIC Class I wells. DWMA applies Web-based mapping and database technology to enhance agency capabilities for identifying major contaminant risks to drinking water sources for both surface water and groundwater. The use of Oracle database systems combined with the mapping capabilities of ArcIMS distinguishes DWMA as one of only a handful of applications at EPA that uses an HTML Web browser to provide a desktop-based GIS system in a single integrated package.

DWMA can generate maps and reports at various scales of analysis using its drill-down, query, analytical, and reporting functions. Basic features include geospatial analysis of water source protection areas (SPAs) and risk analysis using buffer zones around source water intakes and wellheads. DWMA combines multiple query functions and enables spatial queries across multiple data sources, including integration with water quality standards, impaired waters, and many different risk factors. All of these functions are useful in determining program relationships for integration and effectiveness measures.

How DWMA works

DWMA provides an ArcIMS-based viewer tool that enables users to zoom to an area of interest and select from a range of themes to create a high-resolution vicinity map. In most cases, users first access a set of data query panels to define categories of interest. DWMA then applies a drill-down logic to help determine where data meeting the query specifications is available. The drill-down tool gives users the ability to move from national, state, and county level results to more detailed maps simply by clicking on the map itself. Selecting this tool and then clicking on a state or county of interest automatically populates the appropriate query drop-down and initiates the map request. Figure 3.1 demonstrates a sequence of drill-down requests resulting in ever-increasing map resolution.

The primary goal of DWMA is to provide a secure application where EPA staff can easily obtain reports and maps that help manage programs under the SDWA. Some examples of queries that can be addressed in DWMA to generate tables and maps are:

- Query Safe Drinking Water Information System (SDWIS/FED) data on violations of drinking water maximum contaminant levels or other national drinking water regulations
- Display surface water and groundwater (wellhead) protection areas
- Count contaminant risks within SPAs
- Map all available UIC Class I well locations
- Analyze contaminant risks to wellhead protection areas (SPAs around groundwater wellheads) based on geological attributes
- Identify potential contaminant risks to source waters

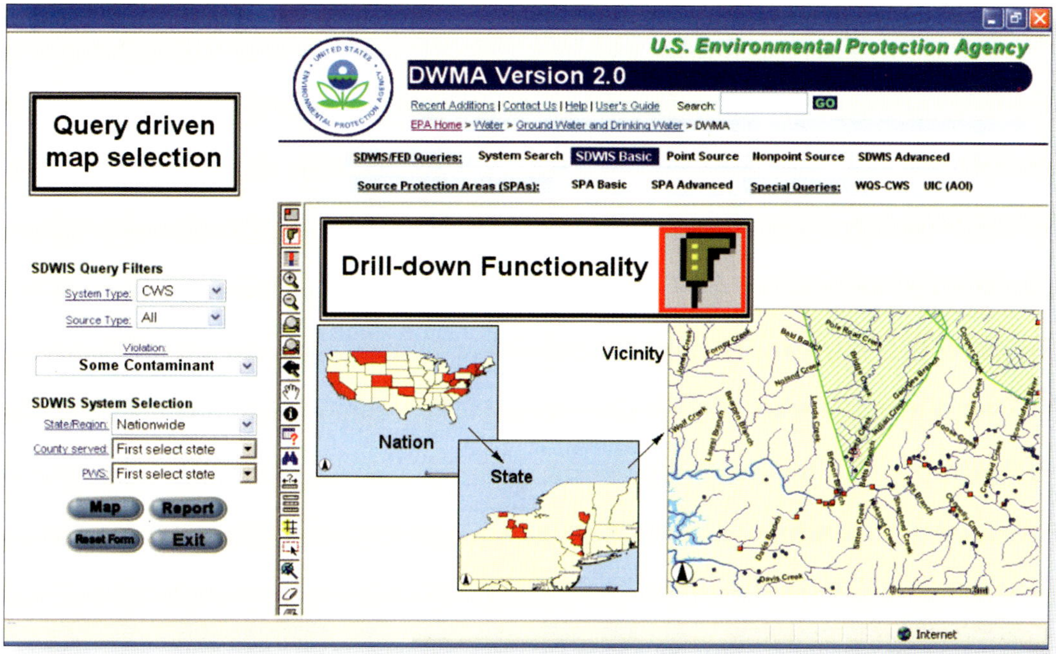

Figure 3.1 DWMA drill-down functionality.

Courtesy of U.S. Environmental Protection Agency.

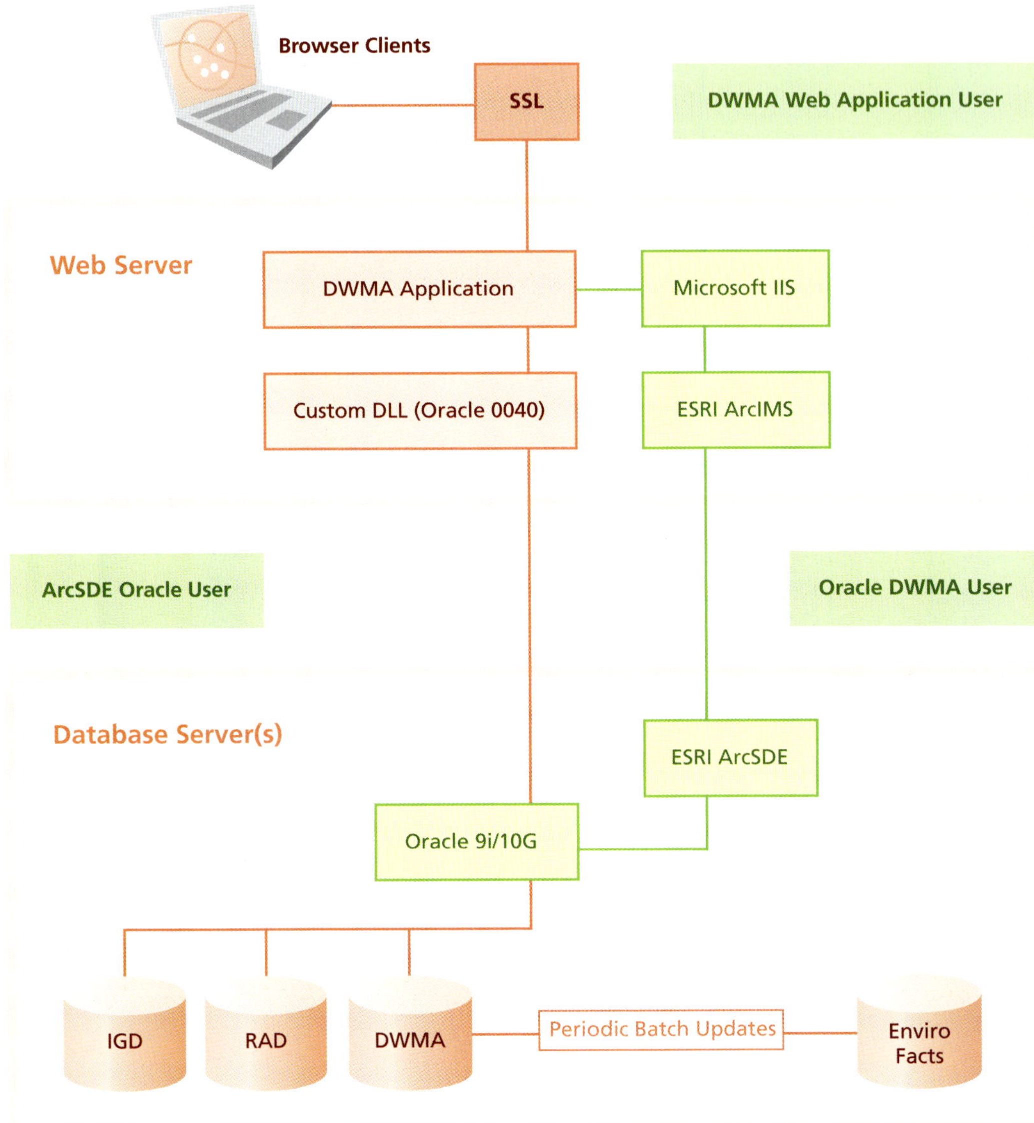

Figure 3.2 DWMA logical design architecture. DLL, Dynamic Link Library; IGD, Integrated GeoDatabase; OO4O, Oracle Objects for Object Linking and Embedding; RAD, Reach Address Database.

Courtesy of U.S. Environmental Protection Agency.

In addition to the USGS NHD data for surface waters, national data sources that are integrated in the application include SDWIS/FED; Permit Compliance System; Resource Conservation and Recovery Act Information System; Comprehensive Environmental Response Compensation and Liability Information System; USGS nitrogen fertilizer use estimates, herbicide use estimates, and karst geology; and USDA land application estimates for animal waste.

Data from the EPA Reach Address Database is based on the NHD, which maintains consistency. Other information in DWMA includes sole source aquifer designations, Department of Defense installations, tribal lands, UIC Class I well locations, hydrogeologic settings and principal aquifers, and surficial and subsurface geology.

The underlying mapping application within DWMA is a customized version of the HTML viewer supplied with ArcIMS. This mapping application is implemented using frames and JavaScript. Other computing technologies used in DWMA include Active Server Pages, Visual Basic Dynamic Link Libraries, Oracle PL/SQL packages and procedures, and Oracle Objects for Object Linking and Embedding. The application requires an Oracle database called DWMA, which consists of a number of Oracle database tables and Oracle packages of PL/SQL procedures. The database tables store relationships between drinking water sources and various potential point and nonpoint sources of contamination. DWMA leverages existing EPA database infrastructure by connecting to the Reach Address Database for NHD and event data and to the EPA Integrated GeoDatabase for other contextual data. While Oracle Spatial is not required to run DWMA, it is used to build and periodically update some of the database tables used by the DWMA. Application access is provided via a unique user and password combination and verified against encrypted values maintained in the DWMA database. The DWMA logical design architecture is summarized in figure 3.2.

Analytical source protection areas

Under the source protection provisions of the Safe Drinking Water Act, EPA is in the process of collecting information from states on source areas for public water supplies. EPA has also implemented procedures to define analytical SPAs around surface intakes and wellheads. Because there are nearly 8,000 surface water intakes associated with public water systems in the United States, the process of identifying analytical SPAs was automated.

The first step in defining the SPAs was to identify upstream drainage networks for each drinking water intake based on medium-resolution NHD (http://nhd.usgs.gov). An upstream distance of fifteen miles was selected as a good approximation of the distance water would travel in a single day. This distance is used by some states for source water assessments, and similar approaches are found in studies of nonpoint source pollution issues. A recursive upstream search algorithm was used to identify upstream drains (such as a small stream segment) within fifteen miles upstream. This procedure was used for each surface water intake.

When the upstream NHD drainage networks were identified, the next step was to build a surface SPA polygon for each network. A goal has been set to define a national set of drain catchments based on methods initially implemented by the USGS for several New England states. After this data is available as part of a system called NHDPlus (a suite of geospatial datasets), drain catchments for the upstream network of drains for a particular drinking water intake can be merged to create the watershed for that intake. Because the identification of these drain catchments was not scheduled to be completed until 2007, analytical watersheds were built using a convex hull approach, which is defined by the set of points that is the smallest convex area that includes all of the drains constituting the upstream network. Figure 3.3 shows an analytical surface polygon defined by the convex hull that surrounds the derived upstream network for an intake point (blue triangle).

SPAs provide a powerful tool for geospatial analysis of information related to vector GIS points, lines, or polygons. They also facilitate processing of raster data layers. For instance, using the 1992 national landcover dataset, impervious-cover percentage metrics were developed for each SPA, providing powerful indicators of landscape units where urbanization processes have altered natural patterns of runoff and infiltration.

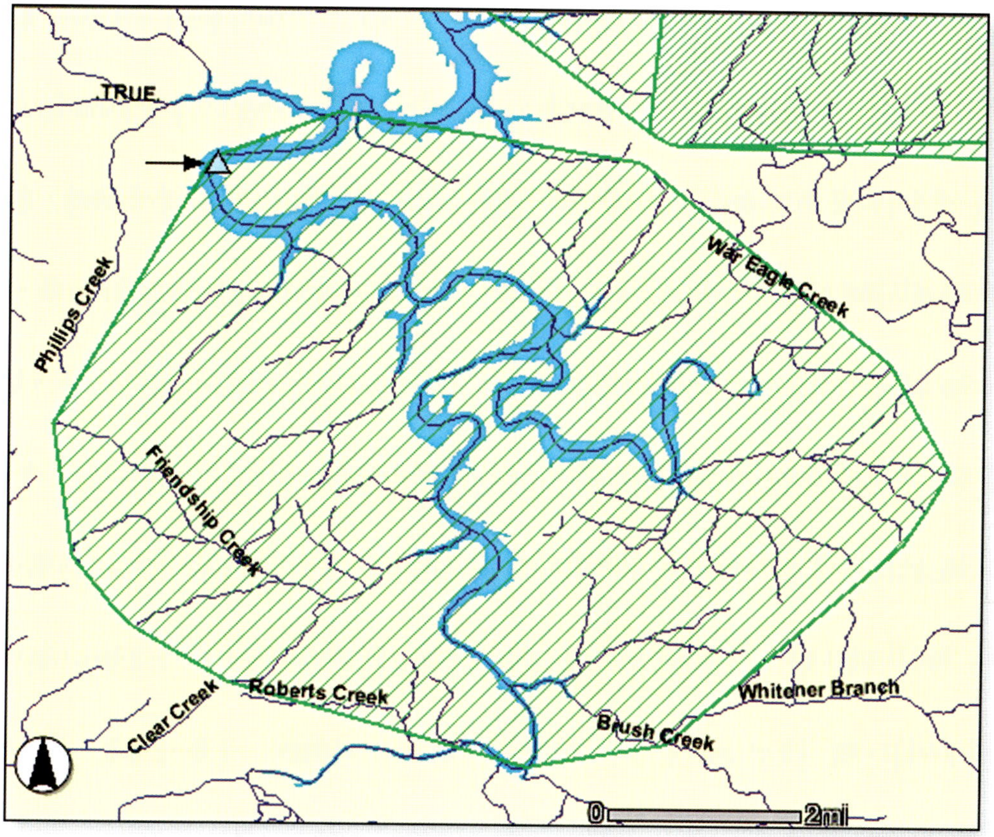

Figure 3.3 Drinking water intake analytical watershed.

Courtesy of U.S. Environmental Protection Agency.

DWMA provides an area-of-investigation query feature that can highlight states or counties containing UIC Class I wells and enables users to search for wellheads within a specified radius of a Class I UIC well (up to ten miles). A particular UIC well location can be investigated with the Wellhead Visualization Tool.

DWMA is available as a secure (password-protected) EPA Intranet tool to designated EPA staff and managers at the national and regional levels for the identification of contaminant risks to surface and groundwater and for protection of public drinking water sources. In the future, DWMA may be expanded to provide secure access to scientists and decision makers at other federal and state natural resource agencies.

OGWDW and other users at EPA headquarters and regional offices have benefited from improved access to information, cost and time savings, and enhanced technical analyses provided by DWMA.

Cooperative management of a protected area

Jeff Nighbert and Kelly Hazen, U.S. Bureau of Land Management, www.blm.gov/or

Steens Mountain Wilderness in Oregon is a remote roadless area containing steep inaccessible terrain. This area is rarely visited and is covered by snow half of the year. Aerial surveys are hazardous and not recommended. Cross-country travel by foot or horseback is the only option. Addressing controversial and high-profile land-use issues in the Steens Mountain area has been difficult, because presenting and describing this unique area to the public is challenging. GIS mapping tools for illustrating elevation range, glacial geology, and viewsheds were not sufficient and did not do the area justice. The new GIS-generated landscape perspectives have given the U.S. Bureau of Land Management (BLM) the visualization tools needed to address the challenges presented by this remote location.

Figure 3.4
Winter scene in Little Blitzen Gorge, taken below Nye Place and looking southwest toward Riddle Brothers Ranch.

Courtesy of U.S. Department of Interior.

The visualizations realistically depict topographic and other natural features, giving land managers a practical alternative to actually visiting the area. This tool is particularly valuable during winter months, when access to many areas is impossible or unsafe.

Figure 3.5 Kiger Gorge is just one of four immense U-shaped gorges formed by glaciers during the Ice Age on Steens Mountain. At over 2,000 feet deep, Kiger exposes lava flows of Steens Basalt and is the only gorge that runs north and south in the area.

Courtesy of U.S. Department of Interior.

Figure 3.6 The Donner und Blitzen River (German for "thunder and lightning") is more than eighty miles of designated wild and scenic river on Steens Mountain. From its headwaters to the Malheur National Wildlife Refuge boundary, this wild river offers scenic glaciated canyons, unique ecosystems, exceptional wild trout fisheries, water for the refuge, and an oasis for waterfowl and other wildlife.

Courtesy of U.S. Department of Interior.

GIS landscape visualization technology has the capability to generate dramatic views of the area that simulate views that would be observed if one were flying over the terrain. These landscape visualizations can be a basis for meaningful public discussion and education. The visualizations realistically depict topographic and other natural features, giving land managers a practical alternative to actually visiting the area. This tool is particularly valuable during winter months, when access to many areas is impossible or unsafe. The maps show several views of major issues, which are important to the effective management of Steens Mountain Cooperative Management and Protection Area.

The main view, Sunrise over the Steens Mountain, Oregon, uses computer technology to enhance the geomorphology of this southeastern Oregon landscape by simulating a low angle over the landscape.

The four additional visualizations illustrate important issues in the management of this area. Computer technology enables GIS technical personnel to generate dramatic views of the area as if one were flying over it in an airplane. This aerial landscape perspective helps managers look at the big picture, which in turn facilitates the development of management actions and the analysis of environmental impacts.

Figure 3.7 This digitally enhanced view of Steens Mountain highlights the dramatic topography and reach of the area. Looking north, the image captures the glacial history and subsequent ruggedness of this beautifully remote corner of southeast Oregon.

Courtesy of U.S. Department of Interior.

The visualizations help communicate to the public the nature and context of land management actions in this area, many of whom have never seen or been to the area but have a stake in the outcome of public policy. The public also uses these three-dimensional perspectives to locate areas of interest and plan their visits. A unique aspect of this visualization project was the collaboration of BLM staff resource specialists and GIS technicians in separate locations on opposite sides of Oregon. This remote collaboration did not hinder the project because it used network capacity and communication skills.

GIS landscape visualization technology allows BLM staff to quickly respond to changes in management direction and emphasis. As land allocation issues change, technicians can easily rotate the angle to enhance the view of the area. They can also view the landscape from angles never used before.

4

Optimum Resource Management

Effectively allocating, analyzing, tracking, and conserving assets is vital for federal agencies to adhere to accountability standards. To ensure adequate funding and expansion of programs, a government agency must lay out a credible plan that supports its decisions and inspires confidence in meeting basic operating requirements. In this regard, GIS technology provides federal agencies with tools for efficient resource management that help to maximize investments, control inventories, and generate revenue or avoid costs.

Databases are expensive to build and maintain. They increase in value when they are shared across departments and between agencies. GIS provides the means for federal agencies to maximize the investment in data—including the hardware, software, and time to collect and manage the data—so that every department head, analyst, and occasional user can take advantage of it. Direct results of the sharing of data and applications include the strengthening of relationships with customers and a large measure of reassurance for the officials holding the purse strings.

As part of standard operating procedures, government agencies are responsible for taking regular inventories of assets, tracking their locations and conditions throughout the lifetime of a project, finding more efficient ways to manage and implement services, and planning for and transitioning to new facilities and sites. By attaching attribute information to the locations of assets and linking appropriate information from other databases, GIS software can easily accommodate these management and control functions.

The incorporation of GIS into the business processes of government allows agencies to more effectively target revenue-generating and cost-avoiding activities. GIS can be used to evaluate property, determine the most efficient routes of travel, identify customers, analyze the effects of or need for services, predict costs, plan construction, and track changes over time. During an emergency, GIS can help responders to quickly assess the situation, see the areas affected, and determine what resources are readily available to deploy.

Managing forest stewardship on private nonindustrial lands

Tim Clark, ESRI Federal Lands and Resources Team; Skip Edel, Colorado State Forest Service; and Jacob Frost, Colorado State Forest Service, www.fs.fed.us/na/sap

U.S. forested land (land that is at least 10 percent covered with trees) is composed of lands managed by federal and state agencies, forest product companies, and private landowners. Of the total forested land, approximately 59 percent is composed of small nonindustrial forests. The U.S. Department of Agriculture Forest Service administers the Forest Stewardship Program, which encourages private forest landowners to manage their lands using professionally prepared forest stewardship plans. These plans address all forest-related resources—such as timber, wildlife, fish, water, recreation, and aesthetics—and integrate them with landowners' objectives.

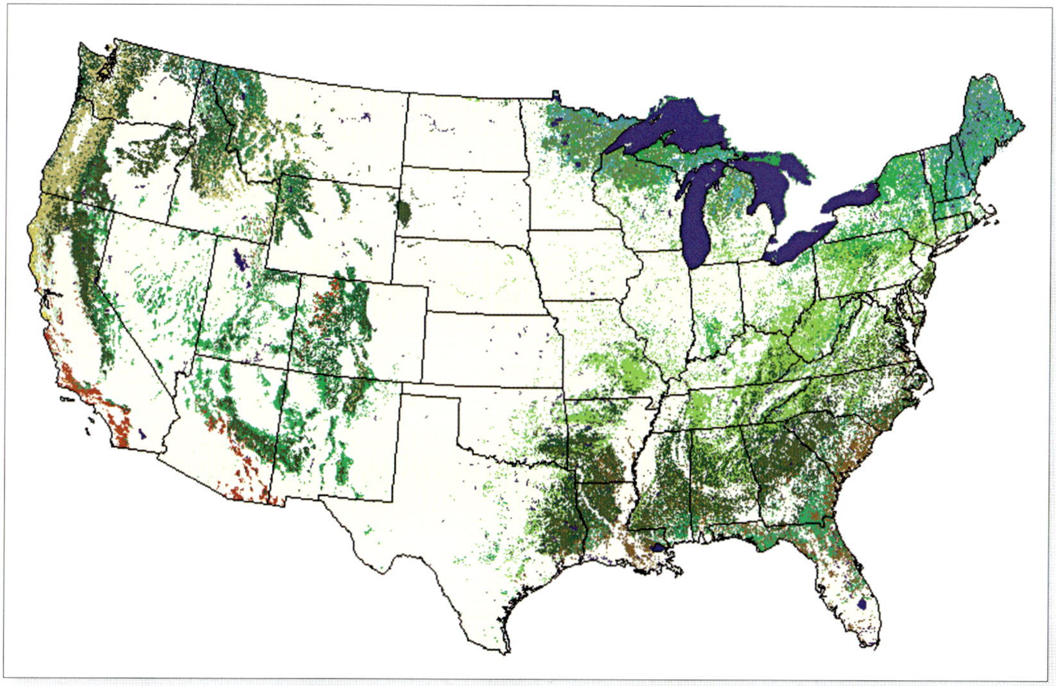

Figure 4.1 Forest resources of the United States.

Courtesy of U.S. Forest Service.

State agency partners have delivered the Forest Stewardship Program since its inception in 1991. Administrators at the state level believed the program had been successful, but it was difficult to demonstrate to Congress and the U.S. Office of Management and Budget that the financial resources invested in the program were being spent the most effective way. This need led to the development of the Spatial Analysis Project (SAP), which enables states to demonstrate prioritization of resources and the effectiveness of resource investments.

Spatial Analysis Project

SAP is a GIS-based strategic management tool that enables participating state forestry agencies to identify and spatially display forest lands that are rich in natural resources and vulnerable to threats. SAP tracks lands currently managed under forest stewardship plans, supports the maintenance of those plans over time, and identifies areas of opportunity for future efforts.

Identifying priority areas for forest stewardship is the first step in the SAP process. This involves collecting twelve layers of information that represent categorizing resource richness (or potential) and resource threats. The nine natural resource richness layers identify private forested lands, wetlands, riparian areas, priority watersheds, forest patches, slope (operability), threatened and endangered species, public water supplies, and proximity to public land. The three resource threat layers show wildfire risk, change in household density, and presence of pathogens.

These twelve layers can vary from state to state; however, the spatial analysis process has the potential to combine layers from neighboring states for analyses of multistate areas. The states determine the value of each layer and its priority ranking for stewardship planning. Using the twelve layers and their relative weights, a geoprocessing model performs a weighted overlay to create an output layer containing a graduated surface of stewardship potential from low to high. Federal lands, bodies of water, and other areas where forest stewardship would not apply are used as masks and excluded from the analysis.

In addition to the twelve core layers, the locations of all existing stewardship plans, tract characteristics, and planned and implemented practices are captured. The existing stewardship plan tracts can then be assessed against the composite stewardship potential surface to quantify the impact of forest stewardship planning and to identify areas of potential opportunities. The final analysis layer is used to address staffing needs that can best focus future stewardship program efforts as well as provide feedback on the effectiveness of the existing program delivery.

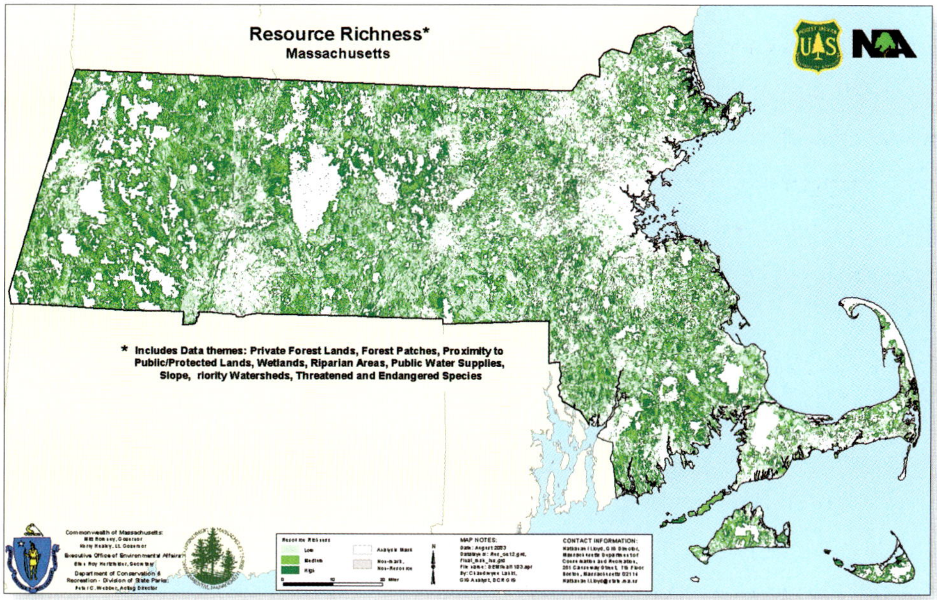

Figure 4.2 Massachusetts natural resource richness.

Courtesy of U.S. Forest Service and the Commonwealth of Massachusetts.

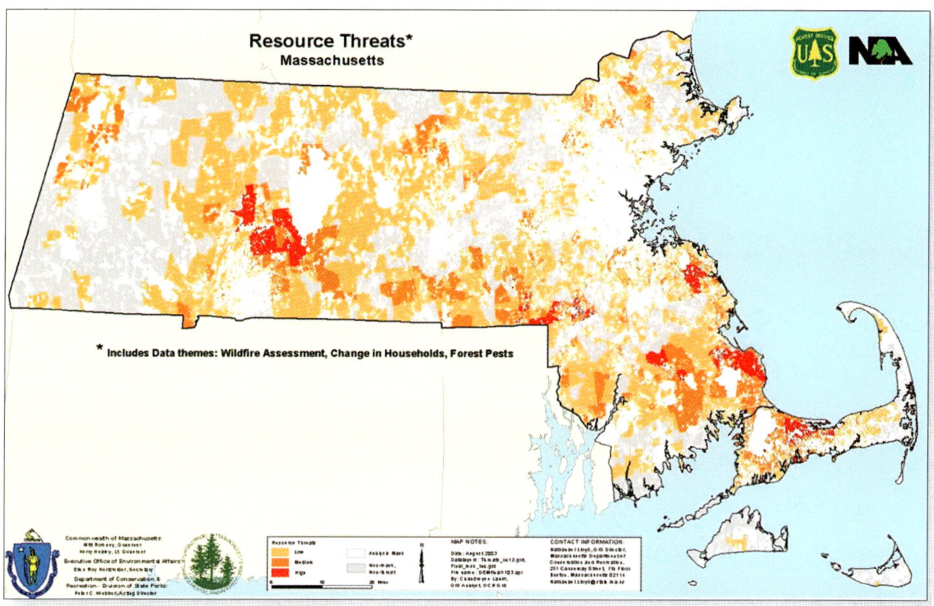

Figure 4.3 Threats to Massachusetts forest resources.

Courtesy of U.S. Forest Service and the Commonwealth of Massachusetts.

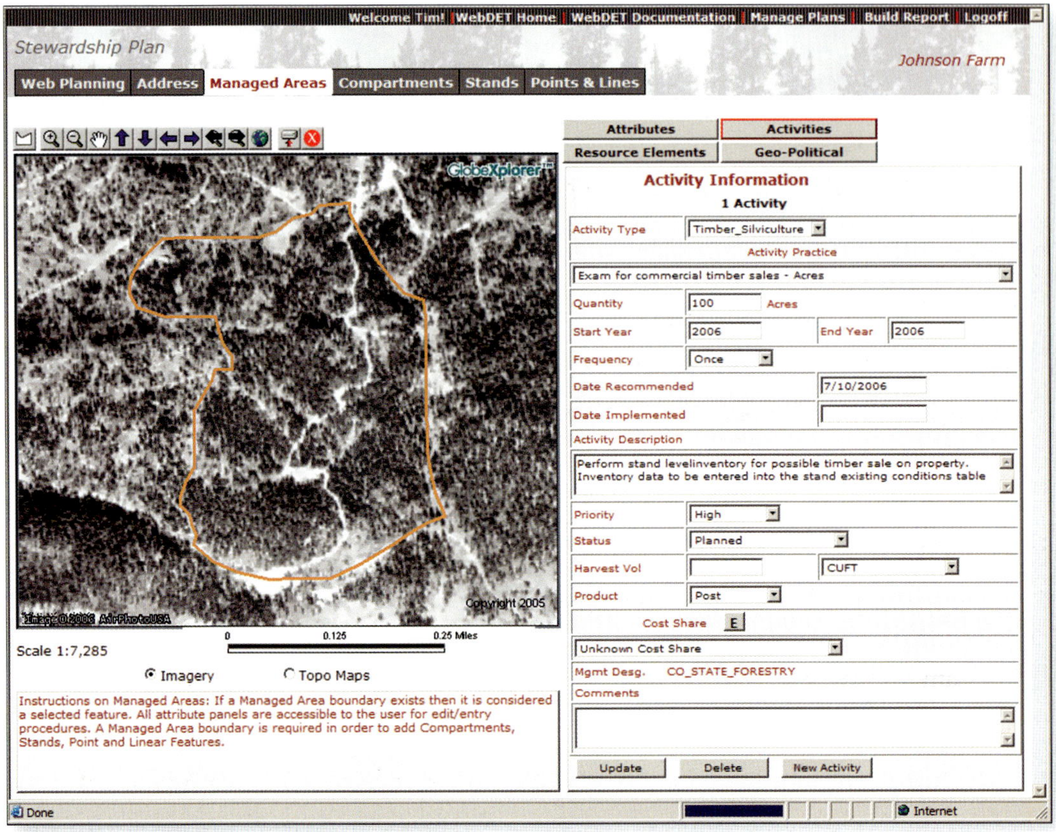

Figure 4.4 Web-DET managed areas activities page.

Courtesy of U.S. Forest Service and the Commonwealth of Massachusetts.

Forest management plans that took many forms and were done manually have been replaced by a standardized map-based digital plan and a comprehensive geodatabase model that supports streamlined reporting as well as ad hoc query capabilities.

Web-DET

After the successful implementation of SAP, states need a mechanism to report program results on a continual basis to optimize strategic program delivery. States need to make the best investment of taxpayer resources without placing onerous reporting demands on field foresters. A reporting tool was needed to help states devise consistent stewardship plans across their agencies.

Representatives from the U.S. Forest Service, state agencies, private forestry companies, and ESRI Federal Land and Natural Resources team assembled in 2003 to determine the business requirements for delivering a forest stewardship plan writer application to state forestry agencies and private consulting foresters across the nation. Key business requirements identified in the needs analysis included ease of use, distribution from a centralized location, a standardized data model, integration of spatial and tabular information into a single tool, and reporting state and federal accomplishments in a timely manner.

It became apparent during the needs assessment that, in addition to generating a forest stewardship plan, the same data model and application could be used to track all field activities of a state forestry agency. The tool also needed to meet the minimum reporting requirements for federal participation and be flexible enough to meet the varying needs of the state agencies.

The business requirements analysis led to the proposal for a Web-based application that would support data entry and editing for the forest stewardship plan and for other state forestry activities. The application would need to accommodate polygon data representing ownership boundaries, resource elements, forest compartments including existing forest stand inventory and desired future conditions, all planned and implemented activities and treatments, and point and line features.

The new stewardship plan writer and data entry tool was named Web-DET (Web-based Data Entry Tool). Web-DET uses ArcGIS Server and the latest technology available to incorporate the plan writer, querying, and reporting in an easy-to-use graphical interface.

Features of Web-DET

- Allows field foresters to enter data and draw maps online
- Integrates a stewardship plan with landowner objectives
- Stores spatial and attribute data in a central database to facilitate reporting
- Accommodates unique state standards and requirements
- Provides an easy-to-use interface for report generation and querying by upper management

Tracking benefits with one tool

As a comprehensive reporting tool, Web-DET facilitates documentation and promotion of program accomplishments and efficient work planning. Its powerful map display capabilities offer a significant advantage over tabular reports and charts.

Web-DET simplifies the reporting process by creating a report that meets the needs of all staff, replacing lengthy or confusing forms. A comprehensive report of accomplishments at the district, state, regional, and national levels can be generated and then viewed many times, greatly improving efficiency. The reporting is as easy as clicking a few buttons, and the report can be saved to disk or sent via e-mail.

Web-DET gives state forestry agencies the ability to efficiently track accomplishments from all programs with one reporting tool. The report shows an aggregate of agency (rather than program) impact and enables state agencies to easily assess how they are reaching their goals and program objectives. Adjustments to program delivery can be made readily, maximizing agency resources. Map-based reporting allows agency staff to visualize program impact over time and across the entire landscape. Map reports are also useful beyond presentation purposes, as they can be used for work planning.

After a SAP statewide assessment is completed and integrated with current stewardship properties, state forestry organizations have the ability to determine which areas would benefit most from implementing the Forest Stewardship Program. This strategic approach to program delivery ensures that resources are invested where they are needed most.

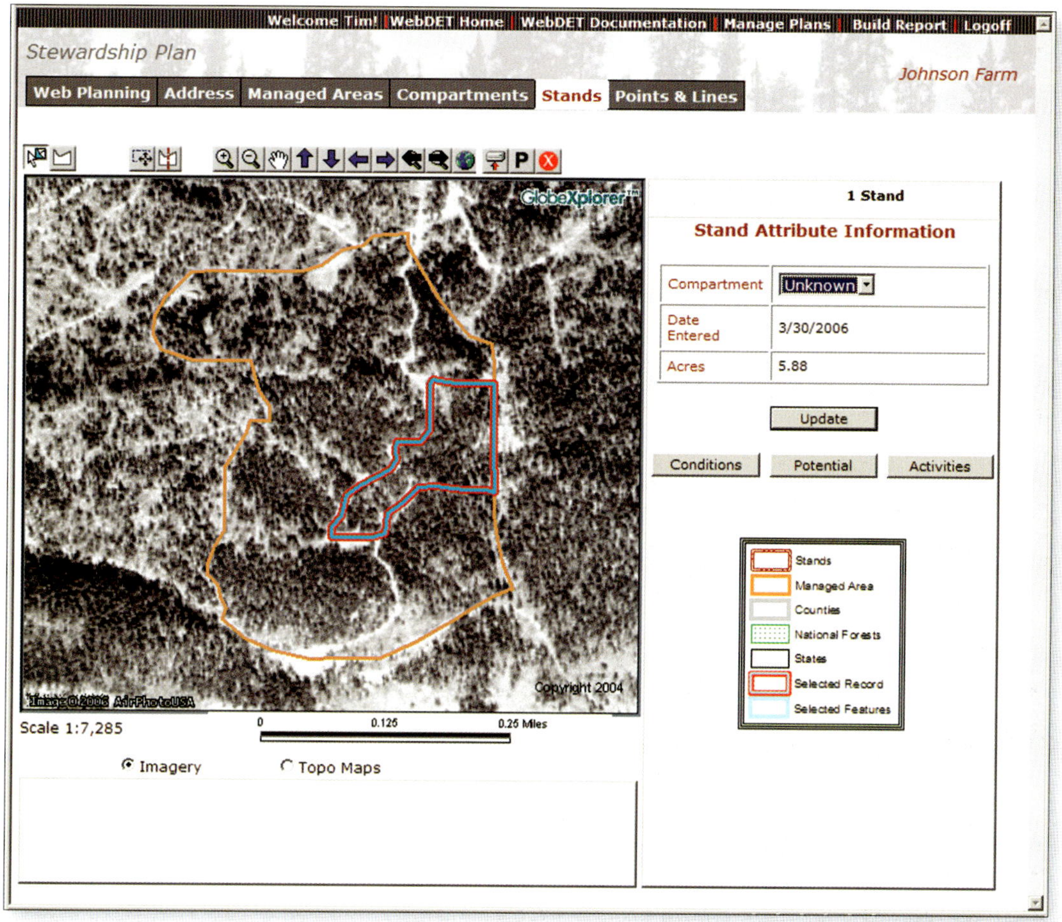

Figure 4.5 Web-DET forest stand attributes page.

Courtesy of U.S. Forestry Service.

A standardized geodatabase model

The standardized geodatabase model has made forest stewardship planning and accomplishment tracking possible at the national level for the first time since the Forest Stewardship Program was started. The geodatabase model enables the deployment of Web-DET and check-in and check-out capabilities for using and editing data locally with ArcGIS. The standards can be integrated with state or regional requirements that are specific to a particular organization without compromising the consistency needed at the federal level. This consistency benefits decision support and analytical processes across the country, as organizations begin to contribute to the enhancement of the standard model.

The next step in the development of Web-DET is to provide Web-DET functionality to those organizations that have little or no access to the Internet. A stand-alone ArcEngine desktop version of Web-DET will provide access to the stewardship geodatabase in a local setting. State offices and other participants will be able to operate with the same set of standards while periodically updating (uploading) their data to the centralized stewardship geodatabase. Occasional access to the Internet will be needed for updating the central geodatabase, but the daily work of data entry, plan preparation, and accomplishment tracking can take place on the local desktop.

SAP and Web-DET have introduced geospatial technologies to meet an existing long-term need. Forest management plans that took many forms and were done manually have been replaced by a standardized map-based digital plan and a comprehensive geodatabase model that supports streamlined reporting as well as ad hoc query capabilities. The U.S. Forest Service can provide this resource to its state customers in an easy-to-use Web-based application. The use of geospatial technologies to analyze forest stewardship priority planning areas also helps to foster collaboration and communication among states as well as between states and the federal government.

Keeping pace with technological innovations: An old facility meets new challenges

William Ball, NASA Langley Research Center

The National Aeronautics and Space Administration (NASA) Langley Research Center (LaRC) is the oldest of ten major NASA centers. Located in Hampton, Virginia, it is adjacent to Langley Air Force Base. LaRC spans 800 acres with 400 structures comprising nearly 7,000 rooms totaling 4 million square feet, and its total assets are valued at $2.4 billion. The facilities were designed to house more than 4,000 civil service and contract employees. LaRC is primarily identified with aeronautical and wind tunnel research, but it supports many other disciplines, including structures and materials, flight electronics, and atmospheric sciences. Such diverse areas of research require a massive and complex infrastructure.

The LaRC GIS team manages plant-level spatial data for the facility and building interior details in GIS. The GIS team consists of three civil servants and five contractors who address the needs of LaRC and outside partners such as NASA headquarters and other NASA centers. Additionally, the team provides periodic support to Air Force bases. The LaRC GIS team has been using spatial data technology for more than a decade and has developed advanced capabilities for LaRC and other organizations it supports.

Doing more with less

LaRC is a complex and old facility (more than fifty years old) with a historic focus on aeronautics. A recent change of emphasis in the NASA mission to space exploration has resulted in downsizing across the agency, specifically at aeronautical facilities. At LaRC, there was a 25 percent reduction in land, buildings, and personnel.

For more than forty years, LaRC infrastructure was reasonably well documented in manually drafted maps and written reports. The paper-based approach became increasingly costly and difficult to maintain. It relied heavily on shrinking reserves of staff members' knowledge and was not effective for analysis or report generation. Staff and budget reductions, along with increasing calls for data by headquarters and LaRC management, created a need for more advanced spatial data processes.

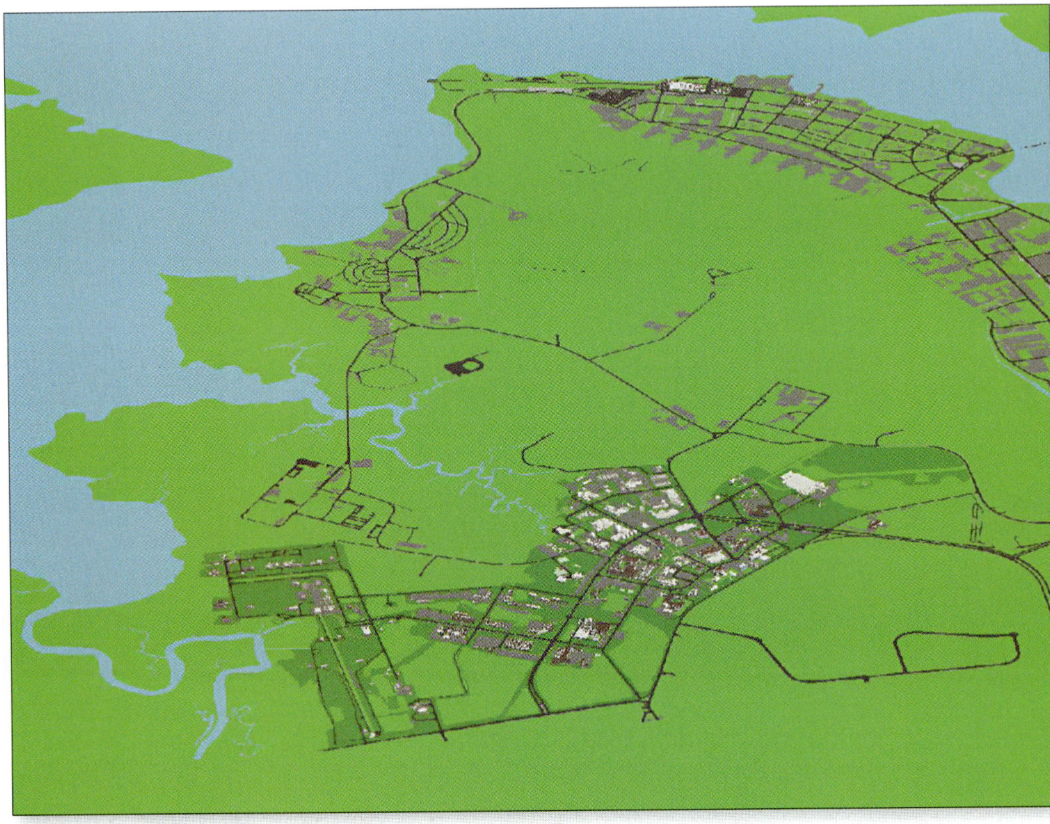

Figure 4.6 LaRC West and East area boundaries relative to adjacent Air Force property.

Courtesy of NASA Langley Research Center.

LaRC has strived to implement state-of-the-art GIS capabilities to address functions such as the master plan, real property, space utilization, facility full cost, underground utilities, building interior details, and environmental issues. The current system consists of ESRI software (ArcInfo, ArcEditor, ArcView, ArcSDE, ArcGIS 3D Analyst, ArcGIS Network Analyst, Image Analysis for ArcGIS, ArcGIS Spatial Analyst, TIN, COGO, GRID, ArcPad, ArcCAD, ArcPress, ArcExpress, ArcScan, MapObjects, ArcStorm, StreetMap, and BusinessMAP), relational database management system software (Oracle), and open-standard, open-source programming languages (Perl, Python, and PHP). The GIS team places a great deal of emphasis on Web-enabled analysis and visualization tools. It is estimated that over 80 percent of analysis occurs via Web tools.

A core function of government is infrastructure maintenance oversight and retention of infrastructure documentation. Accurate capture and maintenance of utility network topology and equipment data can facilitate future outsourcing efforts. To support efforts to reduce the cost of government and to deal with staff reductions, it is imperative to consolidate data handling functions and streamline decision and reporting processes. Besides the economic and functional aspects of maintaining accurate utility and communications information, the government has an inherent responsibility to exercise due diligence in the control of this information as part of homeland security and emergency preparedness.

Improving business processes and decision support

Bottom-line savings are difficult to quantify in government activities, since the "profits" are not generally financial. A government agency such as NASA can strive to minimize operational costs and maximize the ability to serve the public while fulfilling its responsibilities as an employer and member of the community. GIS and LaRC custom tools have greatly improved business processes and decision support capabilities at the center.

Maintaining complete data for systems such as fire protection, drainage, and water facilitates maintenance and fault analysis, both of which can reduce the risk or impact of malfunction and contamination events. Locations for monitoring wells can be optimized based on multidimensional analysis. Network analysis of drain systems combined with a database of locations of hazardous materials can help to trace a chemical spill back to its likely origin.

Complete mapping of Americans with Disabilities Act features and access structures (ramps, elevators) in conjunction with assigned locations of users helps to ensure that adequate accommodations are made for routine access and emergency egress.

The synthesis and analysis capabilities of GIS—backed by utility, topographic, environmental, and personnel databases—enable comprehensive sustainability planning. Optimization of energy use, effluent control and monitoring, and runoff management are all vital parts of this strategy that can benefit from GIS analysis. Responsible environmental management is critical because LaRC is adjacent to wetlands that are directly connected to the Chesapeake Bay.

Up-to-date maps of building layouts with locations of emergency equipment, flammable or toxic storage, high-pressure gases, and other pertinent information are vital to first responders. Locations and characteristics of pressure vessels and flammable and hazardous materials can be incorporated in a study of potential damage from component failure, accident, or malicious acts.

Accurate maps with locations of stored hazardous or flammable materials, valuable components, critical communications trunks, and high-energy systems can help safety teams analyze perimeter security, detect vulnerabilities, design protection strategies, and plan emergency responses. Accurate road information and network analysis can help reduce operating costs for security forces and minimize fire, emergency medical service, and hazardous material transit times by identifying the shortest routes.

The LaRC Flood Tool has proven many times to be an important resource for mission assurance. It has been used repeatedly to help prepare for natural disasters by predicting and displaying flooding at LaRC/Langley Air Force Base based on projected tidal deviations. Users can enter projected tide levels for several local tidal stations from meteorological predictions and display a map of the areas at risk for flooding. Before developing the Flood Tool, the GIS office typically printed 50 D-size maps at a cost of approximately $100 per storm for media alone. In contrast, during Hurricane Isabel in September 2003, 150 people viewed an interactive map online and 10 users printed a map on their local printers, eliminating the need for analysis or printing by the GIS team.

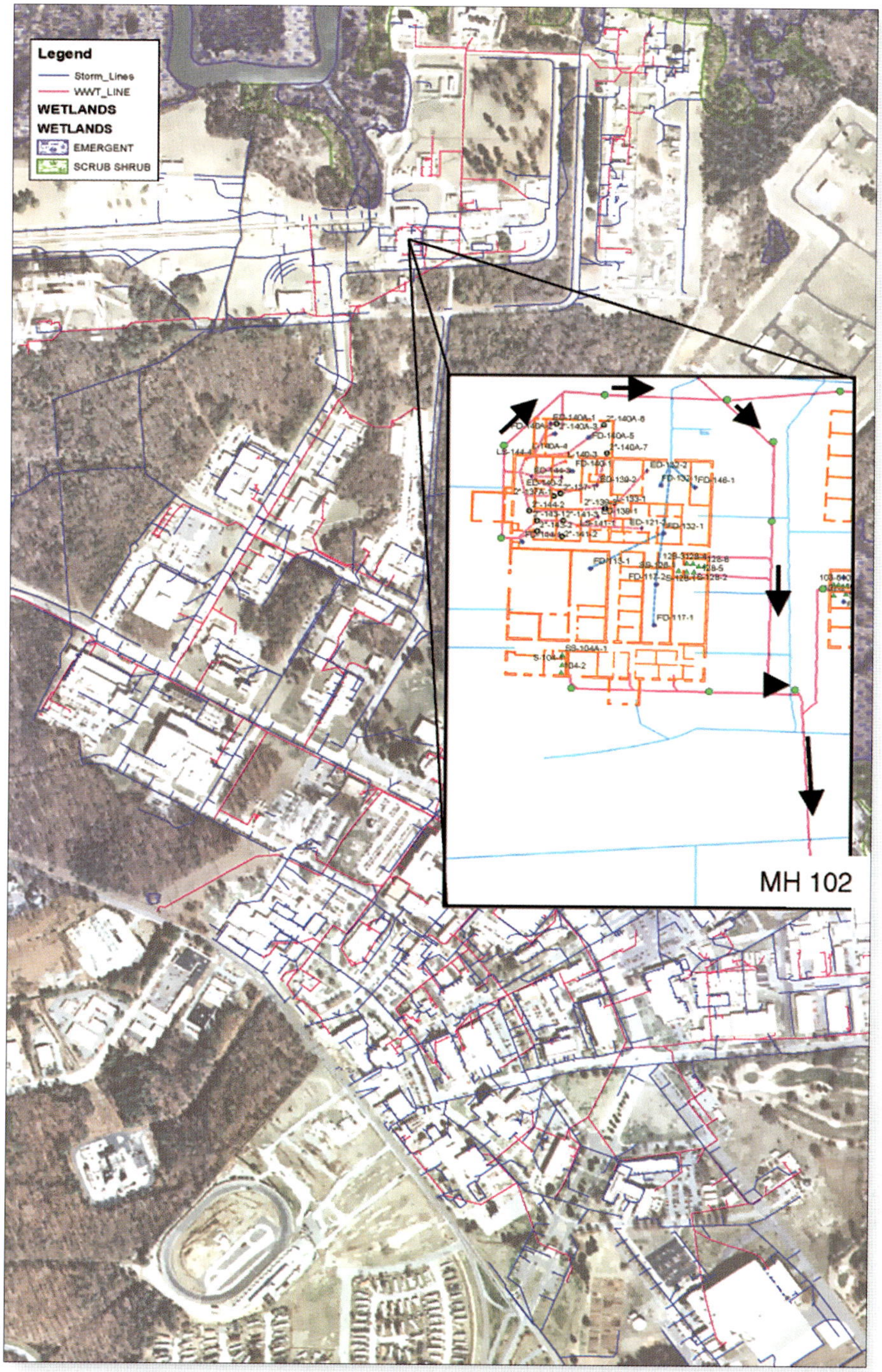

Figure 4.7 Base level sanitary and storm drain system with building interior details.

Courtesy of NASA Langley Research Center.

Figure 4.8 3D extrusions of 2D data based on integrating building- and room-level data with plant-level data.

Courtesy of NASA Langley Research Center.

The Real Property Tool provides important building and facilities information such as property status, square footage, number of occupants, and structure type. Steve Bolman at the maintenance management branch asserts that using the tool is "10 times faster than the old way" of retrieving the data.

Fred Thompson, the space utilization officer from 1998 until 2005, estimates that by using the GIS-based space utilization tool to plan major moves, all of the necessary data can be collected into a report within two weeks. Without the tool, he estimates it would take at least six months to gather the data from the various sources.

The LaRC Facilities Master Plan was revised completely in 2003 to become the single resource linking all spatial data at LaRC. The last formally printed master plan was published in 1993. Distribution was limited to 58 copies, at approximately $150 each. Information was outdated by the time of publication, and the number of printed maps did not meet the demand. The current GIS-based online version of the master plan receives an average of 1,500 visits per month from more than 800 unique customers. To provide these customers with printed copies of the master plan would cost $120,000. Using standard Web browsers, personnel can retrieve information quickly, with the average user viewing six pages in less than five minutes.

Consolidating databases under GIS eliminates redundancy and increases the accuracy and accessibility of information. An excellent example is the use of building floor plans that were maintained by six separate organizations. These plans were important to many departments including those responsible for engineering, maintenance, communications, and space utilization. The plans were updated inconsistently and with varying frequency. Eliminating all but one database provided a single, definitive, consistent set of data for all users. Making the information available on the Web in the form of an intuitive map application with file export and print capabilities allowed a broader audience than just the original few departments to access the data.

Providing a common interface for multiple existing LaRC databases and data storage systems is another feature of GIS and relational database technology. Links to datasets maintained by separate departments allow relationships between those datasets to be explored as an integrated knowledge space. This linking also has cooperative advantages in terms of data quality.

Reaping the benefits

With a better understanding of spatial data, managers can incorporate standards for systems, databases, and documentation into maintenance and construction contracts. Accurate information based on comprehensive GIS data can help contractors bid more accurately and gives government managers the ability to produce better cost and schedule estimates for maintenance work.

GIS improved efficiency in the computation of surface area of roads and parking lots at LaRC for contracted maintenance. Values traditionally maintained in the headquarters (HQ) real property (RP) database were compared to values calculated in GIS, and significant discrepancies were identified:

Method	Streets		Parking	
	Area (sq. yards)	*Difference*	*Area (sq. yards)*	*Difference*
HQ RP	400,000	244.6%	47,000	771.4%
LaRC GIS	978,531		362,545	

Use of optimization algorithms in concert with GIS promises extensive benefits for government facilities, as demonstrated by the following example. The goal of a 2005 office consolidation project at LaRC was to reduce office space utilization from 192 to 125 square feet per person. A manual decision process was used, and the resulting subjective solution moved 2,500 personnel and achieved 149 square feet per person, with an average cost per move of $354 per person. Subsequently, a prototype automated optimization GIS model gave an objective solution whereby the goal of 125 square feet per person could have been achieved by relocating only 1,200 personnel. The optimization approach would have saved the center approximately $460,000, minimized disruption of normal operations, and achieved greater organizational synergy.

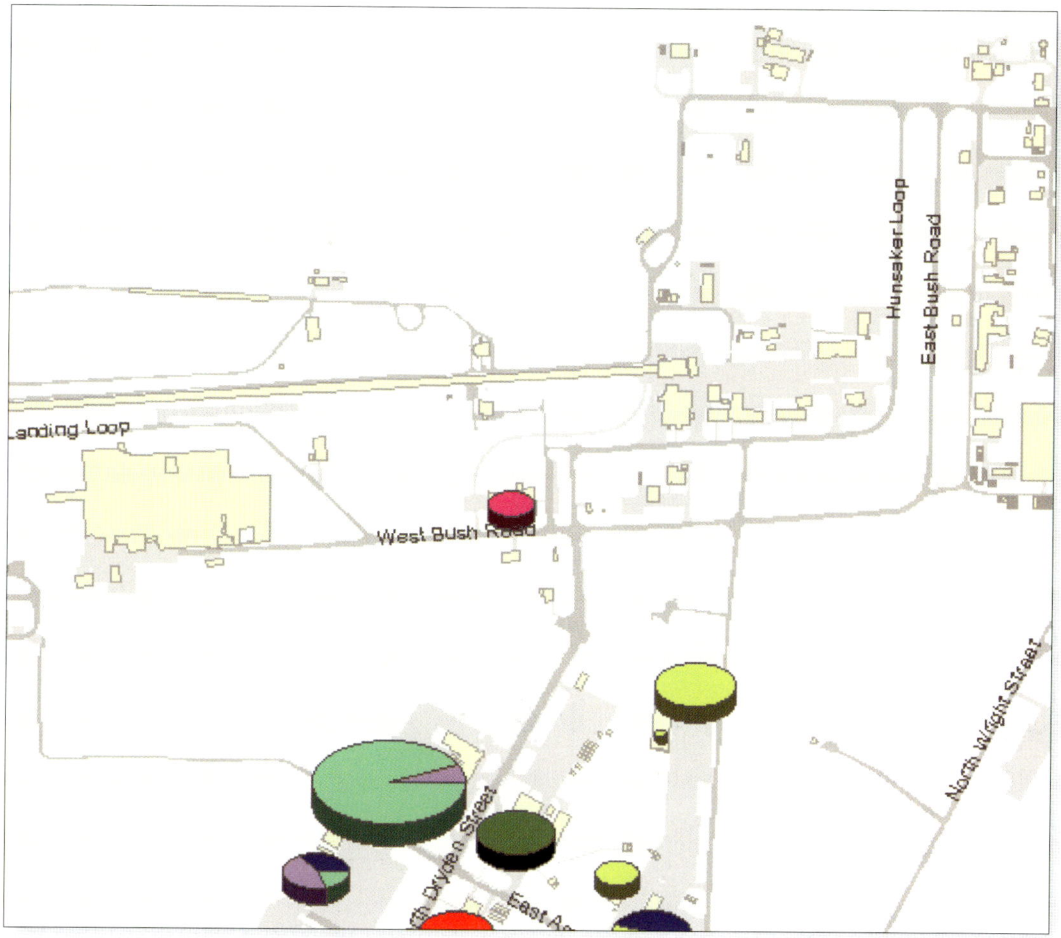

Figure 4.9 Display of building occupancy and organization distribution after use of automated optimization algorithms. Greater density and synergy within organizations is achieved.

Courtesy of NASA Langley Research Center.

The above example addresses only move-related costs for a single effort focused on office space. LaRC has approximately eight times as much space in technical facilities as it has in office facilities, with the current replacement value of the technical facilities dwarfing that of the office facilities approximately thirty-fold. Additionally, the cost for operating technical facilities is nine times that for office facilities. In the LaRC example, operations and maintenance cost avoidance for facilities that have been recently closed is approximately $2.9 million. The use of a GIS-based optimization capability for facility closures at multiple locations would result in hundreds of millions of dollars of savings for the government.

At LaRC, GIS has proven itself a powerful information technology tool. GIS routinely provides faster access to more accurate data than previous approaches. It increases productivity, reduces paperwork, and helps automate many common tasks. It is an intuitive decision-making tool that emphasizes visualization and abstraction from integrated databases. Because of these significant benefits, adoption of GIS for federal facilities is transitioning from a luxury to a necessity.

In the LaRC example, operations and maintenance cost avoidance for facilities that have been recently closed is approximately $2.9 million. The use of a GIS-based optimization capability for facility closures at multiple locations would result in hundreds of millions of dollars of savings for the government.

EPA plans for emergencies with GIS

U.S. Environmental Protection Agency

When a major storm strikes a city and transportation networks are thwarted or shut down, many people, including critical federal personnel, may not be able to reach their primary workplace. In major metropolitan areas such as Philadelphia, a single event can prevent thousands of federal employees from reaching their home office, resulting in significant loss of government services. In a coordinated effort to keep the government working and help agency managers form strategic plans for redeploying personnel to available federal office space, the U.S. Environmental Protection Agency (EPA) and Federal Executive Board (FEB) for the Philadelphia area have begun distributing a GIS-based map to every federal office in the area. The FEB Agency Office Locations map illustrates the distribution of FEB offices in Philadelphia's Center City and areas of Delaware, New Jersey, and Pennsylvania.

FEBs are located in cities that are vital centers of federal activity. These executive boards were established in 1961 by a presidential directive to improve coordination among federal activities and programs outside Washington, D.C. The need for effective coordination among the field activities of federal departments and agencies is clear as approximately 84 percent of all federal employees work outside the national capital area. Along with other functions, FEBs provide a means of communication through which Washington can strengthen field understanding and support management initiatives and concerns. The Strategic Planning and Management Branch of EPA Region 3 has taken the first step in developing a resource that will facilitate redeployment efforts of essential government personnel in response to a natural or human-caused event that limits or prevents employees from reaching their assigned home office.

When certain events occur, such as a major snowstorm, the EPA Region 3 office and other federal agencies in the area can use the map to help redirect employees to auxiliary offices. The map also serves as a central reference point for FEB planners and agency managers.

Continuity of Operations (COOP) is the term the government uses to describe its efforts to maintain essential functions in the event of an emergency. Each federal agency has appointed a COOP planner to develop a plan that details how it will remain operational. The Philadelphia FEB in conjunction with the Federal Emergency Management Agency met with COOP area planners to better plan and coordinate the efforts of federal, state, and local governments in maintaining operations and providing services. The map was developed to aid in this effort and helped identify the federal players.

"Each individual agency in the Philadelphia area has a person designated as the FEB COOP planner," says Michael D'Andrea, program analyst, EPA Region 3. "COOP planners meet regularly and use the FEB Agency Office Locations map to discuss how federal employees will continue to work in the event of an emergency."

Creation of the FEB Agency Office Locations map began with a simple idea to have a reference map showing all FEB offices in the Philadelphia metropolitan region and to include an employee head count for each office. GIS analyst Steve Warner built the initial GIS database and produced cartographic products supporting the basic idea. From there, the concept grew to include a custom symbol set for FEB offices to help indicate a threshold number of employees at a particular office site.

Warner took a list of FEB agency office locations and created a spreadsheet including address, employee count, and agency. He used the GeoCode Service (StreetMap 2003) in ArcMap software to identify locations in the shapefile and manually identified other office locations that were not placed via the GeoCode Service.

Updates are simple, for example, the U.S. Census Bureau office in Center City Philadelphia changed locations. The shapefile was edited by moving the point feature from the old address to the new address (also in Center City) and the attribute table information was changed.

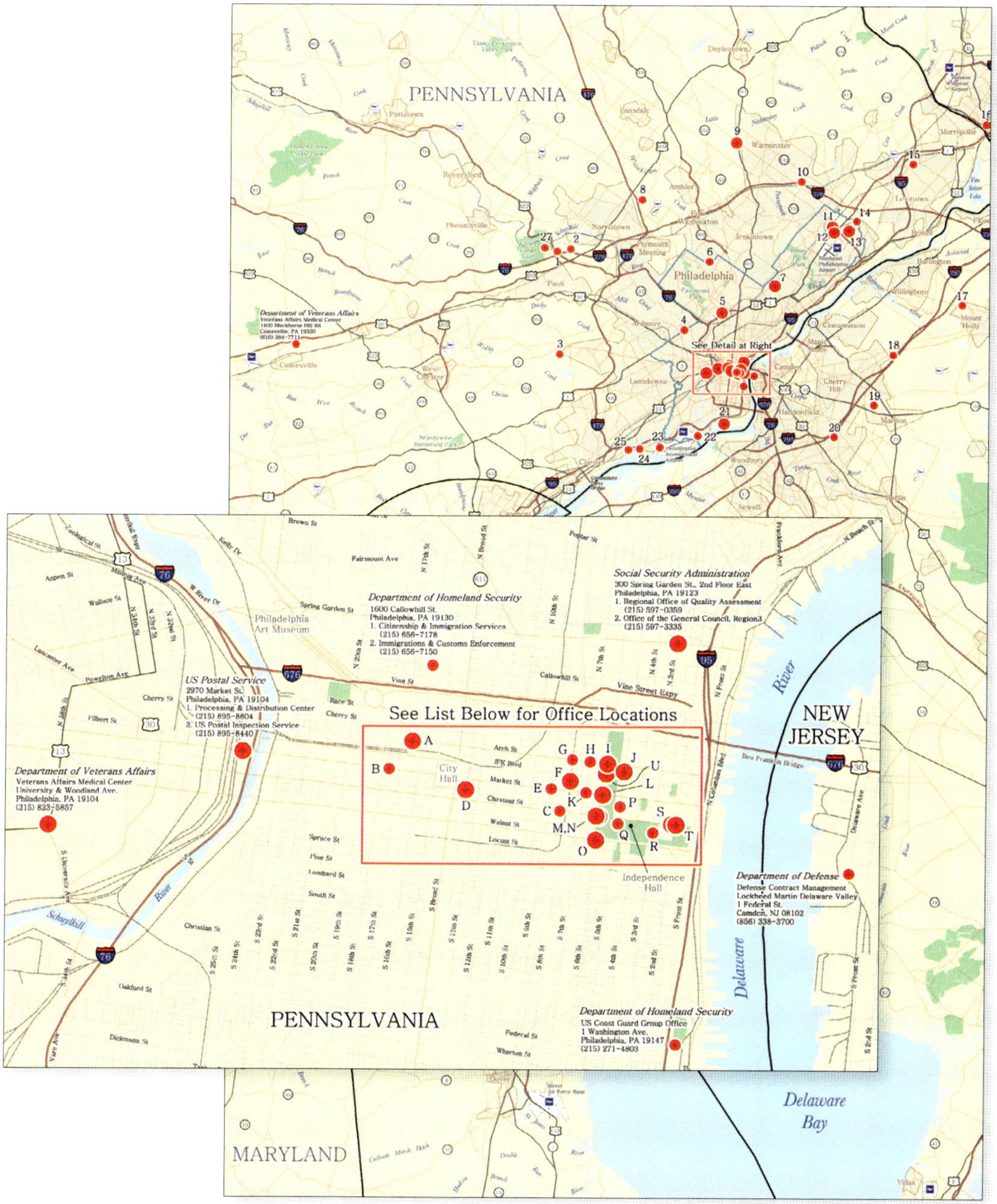

Figure 4.10 The Federal Executive Board Agency Office Locations map is more than a "who's where" of government office locations in and around the Philadelphia metropolitan region. Besides illustrating the distribution of Federal Executive Board offices in Philadelphia's Center City and other areas of Pennsylvania as well as Delaware and New Jersey, this map uses custom symbology to indicate a (number) threshold of employees at an office site.

Courtesy of U.S. Environmental Protection Agency.

Currently, EPA Region 3 staff are updating the FEB Agency Office Locations map to include surface water intakes, reservoirs, transportation hubs, first responder locations, utility grid service areas, and other map layers to help agencies make better decisions. When the map is updated, it will be redistributed to participating FEB offices.

"Eventually," Warner says, "we'd like to build an Internet application using the themes of the map that could be made available to all of the participating FEB agencies." The online application would allow even users unfamiliar with GIS to query the map and perform sophisticated analyses.

Plans call for making the data layers of the map available to individual agencies for their own planning and analysis and for using the digital version of the map as a reference for coordinated planning efforts between participating agencies.

The Strategic Planning and Management Branch of EPA Region 3 has taken the first step in developing a resource that will facilitate redeployment efforts of essential government personnel in response to a natural or human-caused event that limits or prevents employees from reaching their assigned home office.